# VDE VERLAG

Hofheinz, W.

**Vigilancia de la corriente de defecto en instalaciones eléctricas**
Fundamentos, aplicaciones y técnicas de la medición de la corriente diferencial en redes de tensión alterna y continúa IEC 61140, IEC 60364-4-41 con aparatos de vigilancia según, IEC 62020
2005, 134 pages, DIN A5, hardcover
ISBN 3-8007-2834-6
**25.00 €** / 43.80 CHF*

Rittinghaus, D. / Retzlaff, E.

VDE-Schriftenreihe Band 29
**Dictionary of type designation codes for cables and insulated cords according to VDE, CENELEC and IEC**
Lexikon der Kurzzeichen für Kabel und isolierte Leitungen nach VDE, CENELEC und IEC; Deutsch – Englisch
6th, rev. and exp. ed. 2003
208 p., DIN A5, paperback
ISBN 3-8007-2678-5
**22.00 €** / 37.10 CHF*

Werb-Nr. 050815

---

\* Personal VDE members are entitled to 10 % discount.
Orders via booktrade or the publishing house directly.
Prices are subject to change without notice.
The delivery and payment terms of the VDE VERLAG apply to this contract.

Wolfgang Hofheinz

Protective Measures
with Insulation Monitoring

## About the Author

Mr **Wolfgang Hofheinz**, graduate engineer, was born in Wissenbach, Germany. He studied electrical engineering at the University in Giessen, Germany. Since 1975 he is employed by the Bender GmbH & Co. KG in Gruenberg, Germany, becoming the company's Vice-President in 1995. He is honorary expert of various National committees of the DKE (Deutsche Kommission Elektrotechnik Elektronik Informationstechnik) and international electrotechnical standard committees (IEC, NFPA, ASTM). In 2006 he received the "1906 IEC-Award". He has published various expert articles and books on the subject of electrical safety. Some of his books have been translated into English, French, Spanish, Turkish and Hungarian.

Dipl.-Ing. Wolfgang Hofheinz

# Protective Measures with Insulation Monitoring

Application of Unearthed IT Power Systems
in Industry, Mining, Railways,
Marine/Oil and Electric/Rail Vehicles

3rd Edition

**VDE VERLAG GMBH** • Berlin • Offenbach

Cover photo: Bender Pressebild Archiv

Original title:
Schutztechnik mit Isolationsüberwachung, VDE-Schriftenreihe Band 114, 2003
Translated by Monika Patterson

**Bibliografische Information Der Deutschen Bibliothek**
Die Deutsche Bibliothek verzeichnet diese Publikation in der Deutschen
Nationalbibliografie; detaillierte bibliografische Daten sind im Internet über
http://dnb.ddb.de abrufbar

ISBN 978-3-8007-2958-6
ISBN 3-8007-2958-X

© 2006   VDE VERLAG GMBH, Berlin und Offenbach
         Bismarckstraße 33, D-10625 Berlin

All rights reserved

Satz: VDE VERLAG GMBH, Berlin
Druck und Bindung: „Thomas Müntzer" GmbH, Bad Langensalza      2006-10

# Preface to the Third International Edition

The 8th German edition of "Schutztechnik mit Isolationsüberwachung, VDE-Schriftenreihe Band 114, 2003" contained recommendations for the protection against electric shock, with various methods of protection for electrical installations and equipment. The central theme, however, is the "unearthed IT system" (in the following just called IT system) and insulation monitoring.

This translation follows the book at large. However, for the international application, the German and European standard references have been adapted to the respective International standards of the IEC or left out, if not relevant. This necessitated omitting some sections and hence resulted in a different structure of some chapters.

The attentive reader of this 3$^{rd}$ edition will notice that the chapter on IT systems for medical locations is no longer a part of this book. The topic of "electrical safety of medical location" and "safety of medical electrical devices" can now be found in a new German book by the author, titled "Elektrische Sicherheit in medizinischen Bereichen" – the translation "Electrical Safety in Medical Locations" is in planning.

As usual, the rapid technological development and changes in the normative environment makes it almost impossible to publish a book containing the very latest standard updates. However, great effort went into revising the International standards, especially in respect to developments concerning IT systems. Hence it can be said, that the product at hand is more up to date than the last German edition it is based on. And in addition, looking ahead I decided, much against my own principles, to include a few, but very important standard drafts. They are pointed out to the reader.

A new chapter has been added, which had proven to be successful in other International editions. Chapter 17 endeavours to give the reader an overview of the referenced International standards and provides thus a tool for the readers own research.

At this point, I would like to point out, that the European counterparts of the referenced IEC standards in this book may be in another developmental stage and therefore may have a different validity status. This could also mean that European standards (EN – European Norm, or HD – Harmonization Document) may include special guidelines, which are not included in IEC standards, but contain important information for the respective European countries. This development concerns mainly the EN standards: HD 60364-series, EN 60479, HD 61140, described in Chapters 2 to 5 and Chapter 13. It is for this reason, that I am planning to publish an extended International and European edition of this book in 2007, when the development process of these European standards and Harmonization Documents are expected to be completed.

As in the past, criticism, improvements and suggestions are always highly desired, welcomed and, if applicable, considered by me in future publications.

**Special Appreciation**

The research, compilation, verification, collation and summing up of information is tedious and hard to be achieved alone. Writing the original version, I had many helping hands and eyes. A big thank you therefore goes out to:
- Colleagues from various National and International committees
- Colleagues from the BENDER company
- Friends from various sectors of the industry
- Interested readers for their criticism, feed-back and suggestions for improvements.

A very special appreciation is extended to Mrs Monika Patterson, who not only translated this English edition, but at the same time researched International, German and European standards, compiled and collated every chapter and gave many important pointers. Without her support this book would not have been written so quickly.

At this point I explicitly thank the BENDER company in Gruenberg, namely Mr Christian D. Bender, his wife Heidemarie and children Sabine, Christian, Anne and Dorothy, for the exemplary support, not only for writing this book, but for supporting all of the honorary normative work in various national and international standard committees by members of the BENDER-Staff. Their wholehearted belief in the far-reaching benefits of this work to the company is greatly appreciated.

Last but not least my special thanks are also extended to my wife Ellen for her great patience and understanding during my work on this book.

Gruenberg, October 2006                                       Wolfgang Hofheinz

# Preface to the Second Edition

Fundamental changes made by various Standard Organizations as well as the continuous technical development of insulation monitoring devices and insulation fault location systems called for a revision of the first English edition from 1992. In the meantime the 6th edition of the German book has been published. The present edition on the whole is the English translation of the 6th German edition of the book „Schutztechnik mit Isolationsüberwachung". New chapters, such as "Insulation monitoring in rail vehicles" have been included.

The author would like to dedicate special thanks to Mrs. Vollhardt who has done the translation of this book. He also wants to thank Mr. Kaul and Mr. Ernst, Bender Grünberg (Germany), as well as Mr. John Murray, Bender (United Kingdom), who critically examined the draft manuscript.

Grünberg, January 2000                                                       Wolfgang Hofheinz

# Contents

| | | |
|---|---|---|
| 1 | Introduction | 15 |
| 2 | **Protection of Electrical Installations and Equipment** | 19 |
| 2.1 | Common Aspects for Installations and Equipment | 19 |
| 2.1.1 | Scope of IEC 61140 | 19 |
| 2.1.2 | Fundamental Rule of Protection Against Electric Shock | 19 |
| 2.2 | Elements of Protective Measures | 20 |
| 2.2.1 | Basic Protection | 20 |
| 2.2.2 | Fault Protection | 20 |
| 2.2.3 | Enhanced Protective Provisions | 21 |
| 2.3 | Protective Measures | 21 |
| 2.4 | Co-Ordination of Electrical Equipment and of Protective Provisions Within an Electrical Installation | 22 |
| 2.5 | Special Operating and Servicing Conditions | 22 |
| 2.6 | Protective Measures Implementation by Protective Provisions | 23 |
| 3 | **Types of Distribution Systems According to IEC 60364-1** | 25 |
| 3.1 | Types of System Earthing | 26 |
| 3.2 | TN System | 27 |
| 3.2.1 | Single-Source Systems | 27 |
| 3.3 | TT System | 29 |
| 3.4 | IT System | 30 |
| 4 | **Electrical Installations and Protective Measures According to IEC 60364-4-41** | 35 |
| 4.1 | Automatic Disconnection of Supply | 37 |
| 4.1.1 | Residual Current Monitors (RCMs) in Electrical Installations | 37 |
| 4.1.2 | Requirements for Basic Protection | 38 |
| 4.2 | Requirements for Fault Protection | 38 |
| 4.2.1 | Protective Earthing | 38 |
| 4.2.2 | Protective Equipotential Bonding | 38 |
| 4.2.3 | Automatic Disconnection in Case of a Fault | 38 |
| 4.2.3.1 | Disconnection Times for TN and TT Systems | 38 |
| 4.2.3.2 | Disconnection Times for IT Systems in the Event of a Second Fault | 38 |
| 4.2.4 | Additional Protection | 39 |
| 4.3 | TN System | 40 |
| 4.3.1 | Protective Devices in TN Systems | 40 |

| | | |
|---|---|---|
| 4.4 | TT System | 41 |
| 4.4.1 | Protective Devices in TT Systems | 41 |
| 4.5 | IT System | 42 |
| 4.5.1 | Protective Devices and Monitoring Devices in IT Systems | 42 |
| 4.6 | Additional Protective Measures | 44 |
| 4.6.1 | Functional Extra-Low Voltage (FELV) | 44 |
| 4.7 | Protective Measure: Double or Reinforced Insulation | 44 |
| 4.7.1 | Requirements for Basic Protection and Fault Protection | 45 |
| 4.7.1.1 | Electrical Equipment | 45 |
| 4.7.1.2 | Enclosures | 45 |
| 4.7.1.3 | Installation | 45 |
| 4.7.1.4 | Wiring Systems | 46 |
| 4.8 | Protective Measure: Electrical Separation | 46 |
| 4.9 | Protective Measure: Extra-Low-Voltage Provided by SELV and PELV | 46 |
| 4.10 | Additional Protection | 46 |
| 4.10.1 | Additional Protection: Residual Current Protective Devices (RCDs) | 47 |
| 4.10.2 | Additional Protection: Supplementary Equipotential Bonding | 47 |
| 4.11 | Appendices of IEC 60364-4-41 | 48 |
| **5** | **Unearthed IT Systems** | **49** |
| 5.1 | Example of IT Systems with Equipotential Bonding and Insulation Monitoring | 50 |
| 5.2 | Supplementary Protective Equipotential Bonding in the IT System | 53 |
| 5.2.1 | Minimum Cross-Sections for Supplementary Protective Equipotential Bonding | 53 |
| 5.3 | Testing of IT Systems According to IEC 60364-6 | 54 |
| 5.3.1 | Testing | 54 |
| 5.3.2 | Verification for IT Systems | 55 |
| 5.4 | Protection Against Overcurrent in All Distribution Systems | 56 |
| 5.4.1 | Protection of Line Conductors | 56 |
| 5.4.2 | Protection of the Neutral Conductor | 56 |
| 5.4.2.1 | TT or TN Systems | 56 |
| 5.4.3 | IT Systems | 57 |
| 5.4.4 | Protection Against Overload | 57 |
| 5.4.4.1 | Co-Ordination Between Conductors and Overload Protective Devices | 57 |
| 5.4.4.2 | Omission of Devices for Protection Against Overload | 58 |
| 5.4.4.3 | Position or Omission of Devices for Protection Against Overload in IT Systems | 58 |
| 5.4.5 | Cases where Omission of Devices for Overload Protection shall be Considered for Safety Reasons | 59 |
| 5.5 | Connection of Insulation Monitoring Devices (IMDs) | 59 |

| | | |
|---|---|---|
| 5.5.1 | Coupling and Fuse Protection | 60 |
| 5.5.2 | Auxiliary Supply and Fuse Protection | 61 |
| **6** | **Special Features and Advantages of IT Systems** | **63** |
| 6.1 | Higher Operating Safety | 64 |
| 6.2 | Improving Fire Safety | 65 |
| 6.3 | Improving Accident Prevention with Limited Touch Voltages | 67 |
| 6.4 | Higher Permissible Earthing Resistance | 68 |
| 6.5 | Information Advantage with IT Systems | 69 |
| 6.5.1 | Maintenance of Electrical Supply Systems | 71 |
| 6.5.2 | Maintenance Terminology | 72 |
| 6.5.3 | Maintenance Strategy in IT Systems | 72 |
| **7** | **Applications of IT Systems** | **75** |
| 7.1 | IT Systems in the Mining Industry | 75 |
| 7.1.1 | Protective Measures in Underground Mining | 76 |
| 7.1.2 | Standards for Underground Mining | 77 |
| 7.1.2.1 | DIN VDE 0118-1 (VDE 0118-1):2001-11 | 77 |
| 7.1.3 | Protection Against Electric Shock in Underground Mining | 79 |
| 7.2 | IT Systems with Insulation Monitoring on Board Ships | 81 |
| 7.2.1 | Standards and Regulations | 81 |
| 7.2.2 | Permissible System Types Onboard Ships (Marine/Oil) | 82 |
| 7.2.3 | TN and IT Systems on Board Ships | 82 |
| 7.2.4 | IT Systems on Ships of the German Navy (Bundeswehr/Germany) | 85 |
| 7.3 | IT Systems with Insulation Monitoring in Railway Application | 87 |
| 7.3.1 | Examples of Applications for IT System with Insulation Monitoring | 87 |
| 7.3.2 | Application Fields for IT Systems with Insulation Monitoring | 89 |
| 7.3.3 | Requirements on Insulation Monitoring Systems | 89 |
| 7.3.4 | Accummulator-Backed Safety-Oriented d.c. System | 90 |
| 7.3.5 | Converters in Main Circuits | 91 |
| 7.4 | IT Systems in Electric Vehicles | 92 |
| 7.4.1 | Protective Measures in Electric Vehicles | 94 |
| 7.4.2 | Distribution Systems of Electric Vehicles | 95 |
| 7.4.3 | Recharging Stations of Electric Vehicles | 96 |
| 7.4.4 | International Standard for Charging Stations of Electric Vehicles | 98 |
| 7.4.4.1 | Insulation monitoring devices according to UL 2231 | 98 |
| 7.4.4.1.1 | UL 2231-1, General Requirements | 98 |
| 7.4.4.1.2 | UL 2231-2, Supply Circuits | 100 |

| 8 | **Insulation Resistance** | **103** |
|---|---|---|
| 8.1 | Early Safety Regulations in Germany (1883) | 105 |
| 8.2 | Insulation Resistance – a Complex Matter | 106 |
| 8.3 | Definitions | 106 |
| 8.4 | Influence Quantities | 107 |
| 8.5 | Insulation Measurement and Monitoring | 108 |
| 8.5.1 | Measurements in De-Energized Systems | 108 |
| 8.5.2 | Residual Current Monitoring in TN and TT Systems | 108 |
| 8.5.3 | Continuous Monitoring of the Absolute Insulation Value in IT Systems | 109 |
| 8.6 | Complete Monitoring in IT Systems | 110 |
| **9** | **Effects of Shock Current on Humans** | **113** |
| 9.1 | The Effects of Current on Human Beings and Livestock | 114 |
| 9.1.1 | Scope and Object of IEC 60479 | 115 |
| 9.1.2 | Electrical Impedance of the Human Body | 116 |
| 9.1.3 | Sinusoidal alternating Current 50/60 Hz for Large Surface Areas of Contact | 117 |
| 9.1.4 | Threshold of Ventricular Fibrillation | 118 |
| 9.1.5 | Description of Time/Current Zones for a.c. 15 Hz to 100 Hz | 121 |
| 9.2 | Electro-pathological realizations | 121 |
| 9.3 | Protective Measures Against Shock Current | 122 |
| 9.4 | Accidents Involving Electrical Current | 122 |
| **10** | **International Standards for Insulation Monitoring Devices** | **125** |
| 10.1 | Insulation Monitoring Devices (IMD) for Monitoring a.c. Systems in Accordance with IEC 61557-8 | 125 |
| 10.2 | Insulation Monitoring Devices (IMD) in Accordance with IEC 60364-5-53 | 127 |
| 10.3 | Insulation Monitoring Devices (IMD) in Accordance with the American ASTM Standards | 130 |
| 10.3.1 | ASTM F 1207M-96:2002, Standard Specification for Electrical Insulation Monitors for Monitoring Ground Resistance in Active Electrical Systems | 130 |
| 10.3.2 | ASTM F 1669M-96:2002, Standard Specifications for Insulation Monitors for Shipboard Electrical Systems | 130 |
| 10.3.3 | ASTM 1134-94:2002, Standard Specifications for Insulation Resistance Monitor for Shipboard Electrical Motors and Generators | 130 |
| 10.4 | Difference between Insulation Monitoring Devices and Residual Current Monitors in Accordance with IEC 62020 | 131 |
| 10.5 | Equipment for Insulation Fault Location in IT Systems | 131 |

| | | |
|---|---|---|
| **11** | **Technical Implementation of Insulation Monitoring Devices And Earth Fault Monitors** | **135** |
| 11.1 | Insulation Monitoring of a.c. and Three-Phase IT Systems | 135 |
| 11.1.1 | Measurement of Ohmic Insulation Faults | 135 |
| 11.1.2 | Measurement of the Leakage Impedance | 138 |
| 11.2 | a.c. Systems with Directly Connected Rectifiers or Thyristors | 139 |
| 11.2.1 | Measuring with an Inverter | 139 |
| 11.2.2 | Measurement by Pulse Superimposition | 141 |
| 11.3 | d.c. Systems | 142 |
| 11.3.1 | Asymmetric Measurement | 142 |
| 11.3.2 | Measurement by Pulse Superimposition | 143 |
| 11.4 | Measuring Principles for the General Application in a.c. and d.c. IT Systems | 144 |
| 11.4.1 | Microprocessor-Controlled AMP Measurement Process for the General Application in a.c. and d.c. IT Systems | 145 |
| 11.4.2 | Microprocessor-Controlled Frequency-Code Measurement Method for IT Systems with Extreme Interference | 146 |
| 11.5 | Insulation-Fault-Location System in a.c. and d.c. IT Systems | 148 |
| 11.5.1 | Insulation-Fault-Location Systems for d.c. IT Systems | 149 |
| 11.5.2 | Insulation-Fault Location Systems for a.c. and d.c. IT Systems | 149 |
| 11.5.3 | Portable Insulation-Fault-Location System for a.c., d.c. and Three-Phase IT Systems | 150 |
| 11.6 | Summary | 153 |
| **12** | **Response Values of Insulation Monitoring Devices (IMDs)** | **155** |
| **13** | **Physics of the IT System** | **157** |
| 13.1 | Leakage Currents in the IT System | 157 |
| 13.1.1 | Calculation of Leakage Currents in IT Systems | 158 |
| 13.1.2 | Determination of the Leakage Capacitances in the De-Energized System | 158 |
| 13.1.3 | Determination of the Leakage Capacitances in the Energized System | 159 |
| 13.2 | Voltage Ratio in the a.c. IT Systems | 161 |
| 13.3 | Overvoltage in a.c. and Three-Phase a.c. IT Systems | 163 |
| 13.3.1 | Sources of Overvoltage | 163 |
| 13.3.2 | Transient Phenomena at Single-Pole Low Ohmic Insulation Faults | 164 |
| 13.3.3 | Stationary Voltage Rise | 166 |
| 13.3.4 | Intermittent Low Ohmic Insulation Fault | 166 |
| 13.3.5 | Insulation Faults in Supply Systems | 167 |
| 13.3.6 | Switching of Inductivities | 167 |

| | | |
|---|---|---|
| 13.3.7 | Switching of Wires and Capacitors | 168 |
| 13.3.8 | Resonance and Harmonics | 168 |
| 13.3.9 | Voltage Rise at Short-Circuit Disconnection | 168 |
| 13.4 | IT Systems and the Second Fault | 169 |
| 13.4.1 | Fault Constellations in a.c. IT Systems | 171 |
| **14** | **Standard References to IT Systems** | **173** |
| 14.1 | IEC 60364-4-41:2005-12, Low-voltage electrical installations – Part 4-41: Protection for safety – Protection against electric shock | 173 |
| 14.2 | IEC 60364-4-43:2005-07, 64/1557/CDV, Low-voltage electrical installations – Part 4-43: Protection for safety – Protection against overcurrent | 173 |
| 14.3 | IEC 60364-5-53, Ed. 4/CDV:2005-11, IEC 64/1516/CDV, Low-voltage electrical installations – Part 5-53: Selection and erection of electrical equipment – Protection, isolation, switching, control and monitoring | 173 |
| 14.4 | IEC 60364-6:2005, Low-voltage electrical installations – Part 6: Verification | 174 |
| 14.5 | IEC 60364-7-710:2002-11, Electrical installations of buildings – Part 7-710: Requirements for special installations or locations – Medical locations | 174 |
| 14.6 | IEC 60092-507, Ed. 2, IEC 18/1017/CDV:2005, Electrical installations in ships – Part 507: Small vessels | 176 |
| **15** | **Definitions for Insulation Monitoring** | **179** |
| 15.1 | Definitions in accordance with IEC 61557-8:2006 | 179 |
| **16** | **Abbreviations** | **183** |
| **17** | **List of Referenced IEC Standards** | **185** |
| **18** | **Index** | **195** |

# 1 Introduction

Life without electricity is unimaginable. Why certainly! Electrical power supply is taken for granted – electrical and electronic technology is found in all walks of life: private households, industry and commerce, the health system, communication industry or transport to mention but a few. But by the growing demand for electrical devices and equipment, the inherent dangers of electrical power is proportionally posing a threat to people and equipment alike.

The state of the art in the industry for instance, demonstrates the ultimate dependency on a **reliable** power supply. The complexities of electrical installations and their high investment costs, require adequate measures of protection. It is not just the light that goes out, if the power fails. The innumerable headlines of recent years are only too obvious a testimony of the consequences of a power failure: airline crashes, cable car accidents, household fires, computer failure in any branch of industry and commerce, stuck elevators. Just think of the complete power breakdown on the American East coast only recently or the chaos and hardship of the power breakdown in North-West Germany in November 2005. The economic loss alone incurred was immeasurable.

However obvious the direct effects of a power failure, it is often not so obvious to understand, that the real issue is the reliability and safety of the power supply. Safety standards for electrical equipment and installations are therefore of great value. Worldwide there are many standard institutions and committees who are establishing requirements and recommendations in regards to electro-technical issues and who are monitoring, maintaining and upgrading existing standards. The IEC (International Electrotechnical Commission) is one such worldwide organization for standardization comprising of many Technical Committees (TCs). The objective of the IEC is to promote international conformity on all issues concerning standardization in the electrical field.

One important series of IEC standards deals with low-voltage electrical installations, which is under the patronage of the Technical Committee TC 64. It has many sub-committees, each dealing with certain aspects of this important series of IEC 60364.

Since its conception in 1977 the whole series IEC 60364 has been reviewed and restructured several times. New parts are added regularly. Very recently the main title of the whole series has been changed, as seen in Parts 1 and 4-41. **Table 1.1** shows the current status of the series. The new main title will be followed through the other parts in subsequent editions.

| IEC 60364-Part | Main Title<br>Electrical installations of buildings –<br>and title after restructuring*:<br>Low-voltage electrical installations | Edition | Last published |
|---|---|---|---|
| 1 | *Fundamental principles, assessment of general characteristics, definitions | 5.0 | 2005-11 |
| 4-41 | *Protection for safety - protection against electric shock | 5.0 | 2005-12 |
| 4-42 | Protection for safety – protection against thermal shock | 2.0 | 2001-08 |
| 4-43** | Protection for safety - protection against overcurrent | 2.0 | 2001-08 |
| 4-44 | Protection for safety - protection against voltage disturbances and electromagnetic disturbances | 1.1 incl. am1 | 2003-12 |
| 5-51 | *Selection and erection of electrical equipment – Common rules | 5.0 | 2005-04 |
| 5-52 | Selection and erection of electrical equipment – Wiring systems | 2.0 | 2001-08 |
| 5-53** | Selection and erection of electrical equipment – Isolation, switching and control | 3.1 incl. am 1 | 2002-06 |
| 5-54 | Selection and erection of electrical equipment – Earthing arrangements, protective conductors and protective bonding conductors | 2.0 | 2002-06 |
| 5-55 | Selection and erection of electrical equipment – Other equipment | 1.1 incl. am1 | 2002-05 |
| 6 | *Verification | 1.0 | 2006-02 |
| 7-701 | **Requirements for special installations or locations – Locations containing a bath and shower | 2.0 | 2006-02 |
| 7-702 | *Requirements for special installations or locations – Section 702: Swimming pools and other basins | 2.0 | 1997-11 |
| 7-703 | Requirements for special installations or locations – Rooms and cabins containing sauna heaters | 2.0 | 2004-10 |
| 7-704 | *Requirements for special installations or locations – Construction and demolition site installations | 2.0 | 2005-10 |
| 7-705 | Requirements for special installations or locations – Section 705: Electrical installations of agricultural and horticultural premises** | 1.0 | 1984-12 |
| 7-706 | *Requirements for special installations or locations – Conducting locations with restricted movements | 2.0 | 2005-10 |
| 7-708 | Requirements for special installations or locations – Section 708: Electrical installations in caravan parks and caravans | 1.0 incl am1 | 1993-07 |

Table 1.1 Current status of series IEC 60364

| IEC 60364-Part | **Main Title**<br>**Electrical installations of buildings –**<br>and title after restructuring*:<br>**Low-voltage electrical installations** | Edition | Last published |
|---|---|---|---|
| 7-709 | Requirements for special installations or locations – Section 709: Marinas and pleasure craft | 1.0 | 1994-09 |
| 7-710 | Requirements for special installations or locations – Medical locations | 1.0 | 2002-11 |
| 7-711 | Requirements for special installations or locations – Exhibitions, shows and stands | 1.0 | 1998-03 |
| 7-712 | Requirements for special installations or locations – Solar photovoltaic (PV) power supply systems | 1.0 | 2002-05 |
| 7-713 | Requirements for special installations and locations – Section 713: Furniture | 1.0 | 1996-02 |
| 7-714 | Requirements for special installations or locations – Section 714: External lighting installations | 1.0 | 1996-04 |
| 7-715 | Requirements for special installations or locations – Extra-low-voltage lighting installations | 1.0 | 1999-05 |
| 7-717 | Requirements for special installations or locations – Mobile or transportable units | 1.0 | 2001-02 |
| 7-740 | Requirements for special installations or locations – Temporary electrical installations for structures, amusement devices and booths at fairgrounds, amusement parks and circuses | 1.0 | 2000-10 |
| 7-753 | *Requirements for special installations or locations – Floor and ceiling heating systems | 1.0 | 2005-12 |
| * Main title for new editions: Low-voltage electrical installations<br>** Currently under revision | | | |

**Table 1.1** (Continuation) Current status of series IEC 60364

The impetus of restructuring IEC 60364 came from the work of the TCs on the requirements for the erection of low voltage electrical installations in buildings.

Standard IEC 60364-4-41 is of particular importance to the topic of this book and is discussed in detail in Chapter 3.

There are other standards relevant to this book, for instance IEC 60479 "Effects of current on human beings and livestock" see Chapter 8, or IEC 61140 "Protection against electric shock" (see Chapter 2).

In general, when standards are cited in this book the following applies:
- **Dated standards: reference is given only to the standard stated**
- **Undated standards: reference is valid for the latest edition (including all amendments)**

Chapter 17 lists all International Standards and their national equivalent, where applicable.

As a rule, only current standards are referenced in this book, with a few exceptions (**Table 1.2**):

| Standard | Edition | IEC Status | Title | Chapter Reference |
|---|---|---|---|---|
| IEC 60364-4-43 | Ed. 3/CDV: 2006-08 | 64/1557/CDV | Low-voltage electrical installations – Part 4-43: Protection for safety – Protection against overcurrent | 5.4 |
| IEC 60364-5-53 | Ed. 4/CDV: 2005-11<br>RVC: 2006-06 | 64/1516/CDV<br>64/1546/RVC | Low-voltage electrical installations – Part 5-53: Selection and erection of electrical equipment – Protection, isolation, switching, control and monitoring | 10.2 |

Table 1.2 Parts of standard series IEC 60364 under revision, expected to be published soon

The author considers the contents of these standard drafts (CD, CDV), even at this development stage, as important enough to be included in this book. However, they may be subject to further changes by the IEC Technical Committee working groups. The reader is asked, if in doubt, to inquire the status of the above mentioned standard drafts at the IEC Central Office or at their website: www.iec.ch.

# 2 Protection of Electrical Installations and Equipment

## 2.1 Common Aspects for Installations and Equipment

In October 2003 the third edition of the International Standard IEC 61140 was published, with an Amendment in October 2004 titled "Protection against electric shock – Common aspects for installation and equipment". The standard has the status of a basic safety publication intended for use by technical committees in the preparation of new or revising existing standards.

This chapter aims to give the reader an overview of the main statements of this important standard. With the user, the designer and constructor of electrical installations in mind, only the essential parts of the standard are described to keep the information precise. For in-depth-information the interested reader is asked to consult the standard itself (contact details see Chapter 17).

### 2.1.1 Scope of IEC 61140

IEC 61140 applies to the protection of persons and animals against electric shock. It is intended to give fundamental principles and requirements which are common to electrical installations, systems and equipment or necessary for their co-ordination. The standard has been prepared for installations, systems and equipment without a voltage limit. However some clauses refer to low-voltage and high-voltage systems, installations and equipment:

- Low-voltage is any rated voltage up to and including 1000 V a.c. or 1500 V d.c.
- High-voltage is any rated voltage exceeding 1000 V a.c. or 1500 V d.c.

The requirements of the standard apply only if they are incorporated in, or are referred to, the relevant standards. It is not intended to be used as a stand-alone standard.

### 2.1.2 Fundamental Rule of Protection Against Electric Shock

The standard states that hazardous-live-parts shall not be accessible and accessible conductive parts shall not be hazardous live
- either under normal conditions, or
- under single fault conditions.

The accessibility rules of the standard for ordinary persons may differ from those for skilled or instructed persons, and may also vary for different products and locations. For high voltage installations, systems and equipment, entering the danger zone is considered the same as touching hazardous-live-parts.

Protection under normal conditions is provided by basic protection, and protection under single-fault conditions is provided by fault protection. Enhanced protective provisions provide protection under both conditions.

## 2.2 Elements of Protective Measures

All protective provisions shall be designed and constructed to be effective during the anticipated life of the installation, of the system or of the equipment, when used as intended and properly maintained.

The environment should be taken into account by use of the classification of external influences as described in all parts of the series IEC 60721, Classification of environmental conditions. Attention is particularly drawn to the ambient temperature, climatic conditions, presence of water, mechanical stresses, capabilities of persons and area of contact of persons or animals with earth potential.

Fundamentally there are basic, fault and enhanced protective provisions:

### 2.2.1 Basic Protection

Basic protection shall consist of one or more provisions that under normal conditions prevent contact with hazardous-live-parts and are specified as follows:
- basic insulation
- barriers or enclosures
- obstacles
- placing out of arm's reach
- limitation of voltage
- limitation of steady-state touch current and charge
- potential grading

### 2.2.2 Fault Protection

Fault protection shall consist of one or more provisions independent of and additional to those for basic protection, and are specified as follows:
- supplementary insulation
- protective-equipotential-bonding
- protective screening
- indication and disconnection in high-voltage installations and systems
- automatic disconnection of supply

- simple separation (between circuits)
- Non-conducting environment
- potential grading

### 2.2.3 Enhanced Protective Provisions

Enhanced protective provision shall provide both basic and fault protection. Arrangements shall be made so that the protection provided by an enhanced protective provision is unlikely to become degraded and so that a single fault is unlikely to occur. Enhanced provisions are specified as follows:

- reinforced insulation
- protective-separation between circuits
- limited-current-source
- protective impedance device

## 2.3 Protective Measures

This clause in the standard, describes the structure of typical protective measures, indicating in some cases which protective provisions are for basic protection and which are for fault protection.

More than one of the following protective measures may be used within the same installation, system or equipment. Protection by:

- automatic disconnection of supply
- double or reinforced insulation
- equipotential bonding
- electrical separation
- non-conducting environment (low-voltage)
- SELV
- PELV
- limitation of steady-state touch current and charge

## 2.4 Co-Ordination of Electrical Equipment and of Protective Provisions Within an Electrical Installation

Protection is achieved by a combination of the constructional arrangements for the equipment and devices, together with the method of installation.

Equipment may be classified (**Table 2.1**)

- class 0 equipment
- class I equipment
- class II equipment
- class III equipment

Table 2.1 shows the classifications of equipment, according to marking and conditions for connection to the installation.

| Class of equipment | Equipment marking or instructions | Conditions for connection of the equipment to the installation |
|---|---|---|
| Class 0 | – Only for use in non-conductive environment; or<br>– Protected by electrical separation | Non-conducting environment |
| | | Electrical separation provided for each equipment individually |
| Class I | Marking of protective bonding terminal with symbol no. 5019 of IEC 60417-2, or letters PE, or colour combination green-yellow | Connect this terminal to the protective-equipotential-bonding of the installation |
| Class II | Marking with symbol no. 5172 of IEC 60417-2 (double square) | No reliance on installation protective measures |
| Class III | Marking with symbol no. 5180 of IEC 60417-2 (roman numeral III in diamond) | Connect only to SELV and PELV systems |

Table 2.1 Application of equipment in a low-voltage installation according to IEC 61140:2001-10

## 2.5 Special Operating and Servicing Conditions

This clause of IEC 61140 gives detailed requirements for operation of electrical installations and instructions, on the handling of electrical equipment by ordinary persons and skilled or instructed persons, on the location of devices and accessibility.

## 2.6 Protective Measures Implementation by Protective Provisions

Annex A of the standard gives a survey of protective measures as implemented by protective provisions (**Table 2.2**).

|  | **Basic Protection** Protection in absence of a fault | | **Fault Protection** Protection in case of a single fault |
|---|---|---|---|
| Protection by double or reinforced insulation | reinforced insulation | | |
| | Basic insulation Varieties see below | and | Supplementary insulation |
| Protection by equipotential bonding | Basic insulation<br><br>Varieties:<br>• solid basic insulation<br>• basic insulation<br>  – inside barriers and enclosures<br>  – behind obstacles<br>  – placing out of arm's reach | and | Protective-equipotential-bonding Varieties: one or a suitable combination of:<br>• protective-equipotential-bonding (in the installation)<br>• Protective-equipotential-bonding (in equipment)<br>• Protective conductor<br>• PEN-conductor<br>• Protective screening |
| Protection by automatic disconnection of supply | Basic insulation Varieties see above | and | Automatic disconnection of supply |
| Protection by electrical separation | Basic insulation Varieties see above | and | Simple separation (between circuits) |
| Protection by other nonconductive environment | Basic insulation Varieties see above | and | Non conducting environment |
| Protection by other measures | Other provisions | and | Other provisions |
| | Other enhanced protective provisions | | |

**Table 2.2** Protective measures with basic and fault protection

# 3 Types of Distribution Systems According to IEC 60364-1

This chapter describes the systems and their earth connection according to **IEC 60364-1:2005-11, Ed. 5.0, Low-voltage electrical installations – Part 1: Fundamental principles, assessment of general characteristics, definitions.**

This IEC standard is intended to provide safety of persons, livestock and property against danger and damage which may arise in the reasonable use of electrical installations.

The standard gives information on protection against electric shock:

- **as basic protection**

  Protection shall be provided against dangers that may arise from contact with live parts of the installation by persons or livestock.

  This protection can be achieved by one of the following methods:
  - preventing a current from passing through the body of any person or any live stock;
  - limiting the current which can pass through a body to a non-hazardous value.

- **as fault protection**

  Protection shall be provided against dangers that may arise from contact with exposed-conductive-parts of the installation by persons or livestock.

  This protection can be achieved by one of the following methods:
  - preventing a current resulting from a fault from passing through the body of any person or any livestock
  - limiting the magnitude of a current resulting from a fault, which can pass through a body, to a non-hazardous value
  - limiting the duration of a current resulting from a fault, which can pass through a body, to a non-hazardous time period

Essentially the standard identifies three different characteristics of the distribution system:
- types of system earthing
- types of system earthing of the exposed-conductive-parts of the electrical equipment
- characteristics of the protective devices (tripping and alarm devices)

As a result the following characteristics for the type of distribution system are identified:
- type and number of live conductors of the system
- type of system earthing

The systems are distinguished between a.c. and d.c. systems, with different current-carrying conductor systems (**Table 3.1**):

| a.c. systems | d.c. systems |
|---|---|
| Single-phase 2-wire | |
| Single-phase 3-wire | |
| Two-phase 3-wire | 2-wire |
| Two-phase 5-wire | 3-wire |
| Three-phase 3-wire | |
| Three-phase 4-wire | |

Table 3.1 Earthing systems and their allocated conductor systems

## 3.1 Types of System Earthing

Protective measure always require the coordination between types of system earthing and types of protective devices. IEC 60364-1:2005-11, Ed. 5.0, "Low-voltage electrical installations – Part 1: Fundamental principles, assessment of general characteristics, definitions" describe the types of earthing systems as follows.

The various system codes used are derived from the relationship of the distribution system to earth and the relationship of the exposed-conductive-parts of the electrical installation to earth. The codes have the following meaning:

| | |
|---|---|
| **First letter** | – **Relationship of distribution system to earth** |
| T | = direct connection of one point to earth; |
| I | = all live parts isolated from earth or one point connected to earth through an impedance. |
| **Second letter** | – **Relationship of the exposed-conductive-parts of the installation to earth** |
| T | = direct electrical connection of the exposed-conductive-parts to earth, independently of the earthing of any point of the power system; |

| | | |
|---|---|---|
| N | = | direct electrical connection of the exposed-conductive-parts to the earthed point of the power system (in a. c. systems, the earthed point of the power system is normally the neutral point or, if a neutral point is not available, a phase conductor). |
| **Subsequent letter(s)** | – | **Arrangement of neutral and protective conductors** |
| S | = | protective function provided by a conductor separate from the neutral or from the earthed line (or in a. c. systems, earthed phase) conductor |
| C | = | neutral and protective functions combined in a single conductor (PEN conductor) |
| PE | = | protective conductor |

The main distribution systems are:

| TN system | TT system | IT system |
|---|---|---|

The following clauses describe the types of distribution systems according to IEC 60364-1.

## 3.2 TN System

The standard distinguishes the TN systems between single-source systems and multiple source systems.

### 3.2.1 Single-Source Systems

TN distribution systems have one point directly earthed at the source, the exposed-conductive-parts of the installation being connected to that point by protective conductors. Three types of TN systems are considered according to the arrangement of neutral and protective conductors as follows:

- TN-S system in which throughout the system, a separate protective conductor is used (**Figure 3.1**)
- TN-C-S system in which neutral and protective conductor functions are combined in a single conductor in a part of the system
- TN-C system in which neutral and protective conductor functions are combined in a single conductor throughout the system

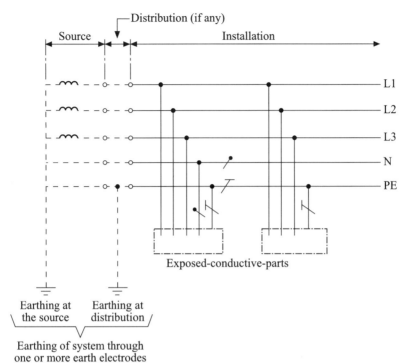

Note: Additional earthing of the PE in the installation may be provided.

**Figure 3.1** TN-S system with separate neutral conductor and protective conductor throughout the system according to IEC 60364-1 [Figure 31A1]

## 3.3　TT System

The TT system has only one point directly earthed and the exposed-conductive-parts of the installation are connected to earth electrodes electrically independent of the earth electrodes of the supply system (**Figure 3.2**).

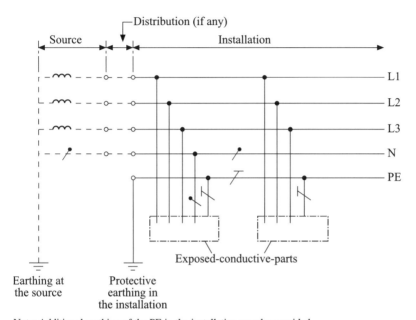

Note: Additional earthing of the PE in the installation may be provided.

**Figure 3.2**　TT System with separate neutral conductor and protective conductor throughout the installation according to IEC 60364-1 [Figure 31F1]

## 3.4 IT System

The IT power system has all live parts isolated from earth or one point connected to earth through an impedance. The exposed-conductive-parts of the electrical installation are earthed
- individually, or
- in groups, or
- collectively (**Figure 3.3**)

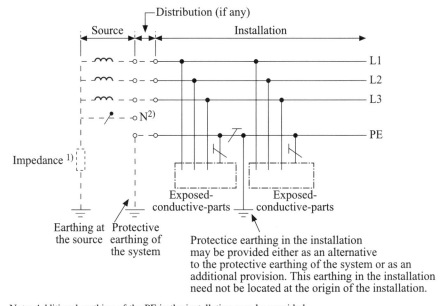

Note: Additional earthing of the PE in the installation may be provided.

1) The system may be connected to earth via a sufficiently high impedance. This connection may be made, for example, at the neutral point, artificial neutral point, or a line conductor.
2) The neutral conductor may or may not be distributed.

**Figure 3.3** IT system with all exposed-conductive-parts interconnected by a protective conductor which is collectively earthed according to IEC 60364-1 [Figure 31G1]

**Figures 3.4, 3.5 and 3.6** show the use of the protective and monitoring devices in the different types of distribution systems.

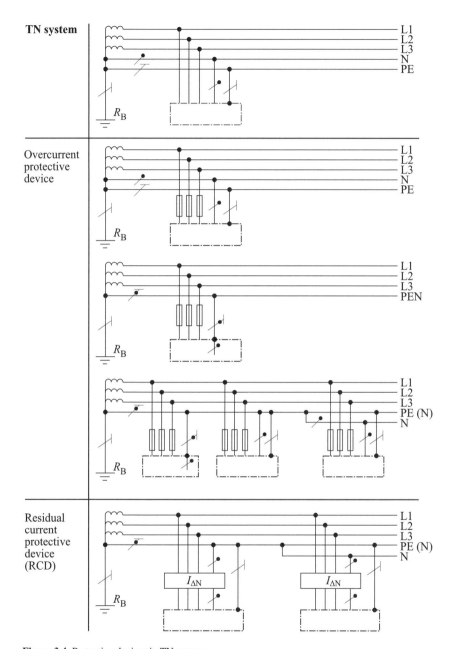

**Figure 3.4** Protective devices in TN systems

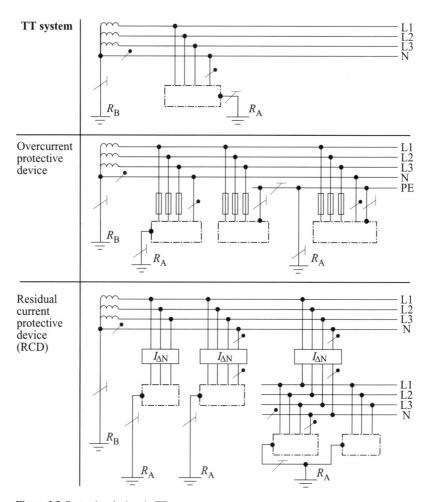

**Figure 3.5** Protective devices in TT systems

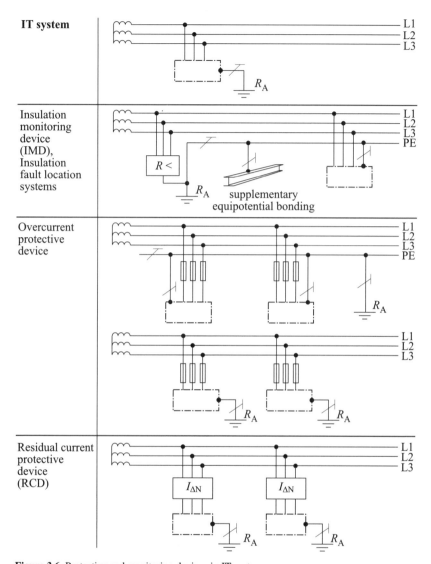

**Figure 3.6** Protective and monitoring devices in IT systems

# 4 Electrical Installations and Protective Measures According to IEC 60364-4-41

The application of electrical energy has been taken for granted in all walks of life. But handling electrical energy naturally places a high demand on the safety of all electrical devices and installations. An electrical installation is usually made up of two sub-systems:

- the supply system and
- the safety system

The supply system is providing electrical energy to users, the purpose of the safety system is to guarantee the protection of persons, animals, installations and equipment.

The International series of standards IEC 60364 with the main title "Low voltage electrical installations" and comprising of many parts, addresses the safety of electrical installations in Part 4, which is sub-titled "Protection for safety" and deals with this subject in four different subchapters, each addressing a different aspects of protection:

| IEC 60364-Part | Low-voltage electrical installations – Subtitle |
| --- | --- |
| 4-41 | Protection against electric shock |
| 4-42 | Protection against thermal effects |
| 4-43 | Protection against overcurrent |
| 4-44 | Protection against voltage disturbancees and electromagnetic disturbances |

The following chapter explains the protection against electric shock with the emphasis on protective measures with insulation monitoring in unearthed IT systems and is based on **IEC 60364-4-41, 5th edition:2005-12[1], Low-voltage electrical installations – Part 4-41: Protection for safety – Protection against electric shock**.

---

1 The original title of the standard "Electrical installations of buildings" has been changed (see Chapter 1). Since this is a brand-new revision of the standard, at the revising of this book, the author considers it interesting to list the countries, which in November 2005 have voted in favour of Part 4-41 of this series of standards: Australia, Belgium, Brazil, Canada, China, Czech Republic, Denmark, Finland, France, Germany, Greece, Hungary, Indonesia, Ireland, Italy, Japan, Korea (Rep. of), Lithuania, Luxembourg, Malta, Mexico, Netherlands, New Zealand, Norway, Poland, Romania, Russian Fed., Slovenia, South Africa, Spain, Sweden, Switzerland, U.S.A, United Kingdom

Amongst other changes to the previous edition, edition 5 considers IT systems more fully.

The fundamental rule of protection against electric shock is that hazardous-live-parts must not be accessible and accessible conductive parts must not be hazardous live, neither under normal conditions nor under single fault conditions.

Protection under normal conditions is provided by **basic protective provisions** and protection under **single fault conditions** is provided by fault protective provisions. In the previous fourth edition, protection under normal conditions (now designated as basic protection) was referred to as protection against direct contact. Protection under fault conditions (now designated as fault protection) was referred to as protection against indirect contact.

The scope of IEC 60364-4-41 states that essential requirements regarding protection against electric shock, including basic protection and fault protection of persons and live-stock are specified. It deals also with the application and co-ordination of these requirements in relation to external influences.

In general a protective measure shall consist of

- an appropriate combination of a provision for basic protection and an independent provision for fault protection, or
- an enhanced protective provision which provides both basic protection and fault protection.

In each part of an installation one or more protective measures shall be applied, taking into account conditions of external influence. Generally permitted are:

- automatic disconnection of supply
- double or reinforced insulation
- electrical separation for the supply of one item of current-using equipment
- extra-low voltages.

The different protective measures are described in the following sub-clauses.

## 4.1 Automatic Disconnection of Supply

In general automatic disconnection of supply is a protective measure in which
- basic protection is provided by basic insulation of live parts or by barriers or enclosures, and
- fault protection by protective equipotential bonding and automatic disconnection in case of a fault.

Where specified, additional protection is provided by a residual current protective device (RCD) with rated residual operating current not exceeding 30 mA.

### 4.1.1 Residual Current Monitors (RCMs) in Electrical Installations

For the first time information is given for the use of residual current monitors (RCMs) in electrical installations in 411.1, Note 2 of IEC 60364-4-41:

Residual current monitors (RCMs) are not protective devices but they may be used to monitor residual currents in electrical installations. RCMs produce an audible or audible and visual signal when a preselected value of residual current is exceeded.

**Figure 4.1** shows a residual current monitor (RCM). **Figure 4.2** shows a series of current transformers.

**Figure 4.1** Residual current monitor type RCM420 (Photo supplied by Bender, Gruenberg, Germany)

**Figure 4.2** Series of current transformer for RCMs (Photo supplied by Bender, Gruenberg, Germany)

### 4.1.2 Requirements for Basic Protection

All electrical equipment shall comply with one of the provisions for basic protection described in Annex A, or where appropriate in Annex B of IEC 60364-4-41.

## 4.2 Requirements for Fault Protection

The following clauses describe the different possibilities for fault protection.

### 4.2.1 Protective Earthing

Exposed-conductive-parts shall be connected to a protective conductor under the specific conditions for each type of system earthing.

### 4.2.2 Protective Equipotential Bonding

In each building the earthing conductor, the main earthing terminal and the following conductive parts shall be connected to the protective equipotential bonding:
- metallic pipes supplying services into the building, e. g. gas, water
- structural extraneous-conductive-parts if accessible in normal use, metallic central heating and air-conditioning systems
- metallic reinforcements of constructional reinforced concrete, if reasonably practicable

### 4.2.3 Automatic Disconnection in Case of a Fault

A protective device shall automatically interrupt the supply to the line conductor of a circuit in the event of a fault of negligible impedance between the line conductor and an exposed-conductive-part or a protective conductor in the circuit or equipment within the disconnection time.

**For IT systems, automatic disconnection is not usually required on the occurrence of a first fault.**

#### 4.2.3.1 Disconnection Times for TN and TT Systems

The maximum disconnection times for final circuits not exceeding 32 A are given in **Table 4.1**.
- In TN systems, a disconnection time not exceeding 5 seconds is permitted for distribution circuits; in TT systems it is 1 second.
- If automatic disconnection cannot be achieved in the required time, supplementary equipotential bonding shall be provided.

| System | 50 V < $U_0$ ≤ 120 V | | 120 V < $U_0$ ≤ 230 V | | 230 V < $U_0$ ≤ 400 V | | $U_0$ > 400 V | |
|---|---|---|---|---|---|---|---|---|
| | s (Seconds) | | s (Seconds) | | s (Seconds) | | s (Seconds) | |
| | a.c. | d.c. | a.c. | d.c. | a.c. | d.c. | a.c. | d.c. |
| TN | 0.8 | Note 1 | 0.4 | 5 | 0.2 | 0.4 | 0.1 | 0.1 |
| TT | 0.3 | Note 1 | 0.2 | 0.4 | 0.07 | 0.2 | 0.04 | 0.1 |

$U_0$ is the normal a.c. or d.c. line to earth voltage

Note 1 Disconnection may be required for reasons other than protection against electric shock.

**Table 4.1** Maximum disconnection times according to IEC 60364-4-41, Table 41.1

### 4.2.3.2 Disconnection Times for IT Systems in the Event of a Second Fault

After the occurrence of a first insulation fault, conditions for automatic disconnection of supply in the event of a second insulation fault occurring on a different live conductor shall be as follows:

- Where exposed-conductive-parts are interconnected by a protective conductor collectively earthed to the same earthing system, the time stated in Table 4.1 for TN systems is applicable, independent if the neutral conductor is distributed or not.
- Where exposed-conductive-parts are earthed in groups or individually, the time stated in Table 4.1 for TT system is applicable. The residual current should typically be 5 times of the rated residual current $I_{\Delta n}$ of the RCD.

### 4.2.4 Additional Protection

In a.c. systems, additional protection by means of a residual current protective device (RCD) shall be provided for

- socket-outlets with a rated current not exceeding 20 A that are for use by ordinary persons and are intended for general use
- mobile equipment with a current rating not exceeding 32 A for use outdoors

## 4.3  TN System

In TN systems the integrity of the earthing of the installation depends on the reliable and effective connection of the PEN or PE conductors to earth. Where the earthing is provided from a public or other supply system, compliance with the necessary conditions external to the installation is the responsibility of the supply network operator.

The neutral point or the midpoint of the supply system shall be earthed. If a neutral point or midpoint is not available or not accessible, a line conductor shall be earthed.

In fixed installations, a single conductor may serve both as a protective conductor and as a neutral conductor (PEN conductor). No switching or isolating device shall be inserted in the PEN conductor.

The characteristics of the protective device and the circuit impedances shall fulfil the following requirement:

$Z_s \times I_a \leq U_0$

where

$Z_s$  is the impedance in ohms ($\Omega$) of the fault loop comprising
 – the source
 – the line conductor up to the point of the fault, and
 – the protective conductor between the point of the fault and the source

$I_a$  is the current in amperes (A) causing the automatic operation of the disconnecting device; when a RCD is used, this current is the residual operating current providing disconnection

$U_0$  is the nominal a. c. or d. c. line to earth voltage in volts (V)

### 4.3.1  Protective Devices in TN Systems

The following protective devices may be used for fault protection in TN systems:

- overcurrent protective devices
- residual current protective devices (RCDs)

## 4.4 TT System

All exposed-conductive-parts collectively protected by the same protective device shall be connected by the protective conductors to an earth electrode common to all those parts. Where several protective devices are utilized in series, this requirement applies separately to all the exposed-conductive-parts protected by each device.

Generally in TT systems, RCDs shall be used for fault protection. Alternatively, overcurrent protective devices may be used for fault protection, provided a suitably low value of $Z_S$ is permanently and reliably assured.

Where a RCD is used for fault protection the circuit should also be protected by an overcurrent protective device the following conditions shall be fulfilled:

$$R_A \times I_{\Delta n} \leq 50 \text{ V}$$

where

$R_A$ is the sum of the resistance in $\Omega$ of the earth electrode and the protective conductor for the exposed-conductive-parts

$I_{\Delta n}$ is the rated residual operating current of the RCD

Where an overcurrent protective device is used for fault protection, the following conditions shall be fulfilled:

$$Z_S \times I_a \leq U_0$$

where

$Z_S$ is the impedance in $\Omega$ of the fault loop comprising of
  – the source
  – the line conductor up to the point of the fault
  – the protective conductor of the exposed-conductive-parts
  – the earthing conductor
  – the earth electrode of the installation and
  – the earth electrode of the source

$I_a$ is the current in A causing the automatic operation of the disconnecting device

$U_0$ is the nominal a.c. or d.c. line to earth voltage

### 4.4.1 Protective Devices in TT Systems

The following protective devices may be used for fault protection in TT systems:
- generally residual current protective devices (RCDs) shall be used
- alternatively overcurrent protective devices may be used by low value of $Z_S$

## 4.5 IT System

The IT system has all live parts isolated from earth or one point connected to earth through a sufficiently high impedance. This connection may be made either at the neutral point or mid-point of the system or at an artificial neutral point. The latter may be connected directly to earth if the resulting impedance to earth is sufficiently high at the system frequency. Where no neutral point or mid-point exists, a line conductor may be connected to earth through a high impedance.

The fault current is then low in the event of a single fault to exposed-conductive-part or to earth and automatic disconnection is not imperative. Provisions shall be taken, however, to avoid risk of harmful pathophysiological effects on a person in contact with simultaneously accessible exposed-conductive-parts in the event of two faults existing simultaneously.

Exposed-conductive-parts shall be earthed individually, in groups, or collectively.

The following condition shall be fulfilled:

- in a.c. systems $\quad R_A \times I_d \leq 50$ V
- in d.c. systems $\quad R_A \times I_d \leq 120$ V

where

$R_A$    is the sum of the resistance in $\Omega$ of the earth electrode and the protective conductor for the exposed-conductive-parts

$I_d$    is the fault current in A of the first fault of negligible impedance between a line conductor and an exposed-conductive-part; the value of $I_d$ takes account of leakage current and the total earthing impedance of the electrical installation.

### 4.5.1 Protective Devices and Monitoring Devices in IT Systems

In IT systems the following monitoring devices and protective devices may be used:

- insulation monitoring devices (IMDs)
- residual current monitoring devices (RCMs)
- insulation fault location systems
- overcurrent protective devices
- residual current protective devices (RCDs)

*Note: Where a residual current protective device (RCD) is used, tripping of the RCD in event of a first fault cannot be excluded due to capacitive leakage currents.*

In cases where an IT system is used for reasons of continuity of supply, an insulation monitoring device (IMD) shall be provided to indicate the occurrence of a first fault from a live part to exposed-conductive-parts or to earth. This device shall initiate an audible and/or visual signal which shall continue as long as the fault persists.

If there are both audible and visible signals, it is permissible for the audible signal to be cancelled.

*Note: It is recommended that a first fault be eliminated with the shortest practical delay.*

Except where a protective device is installed to interrupt the supply in the event of the first earth fault, a RCM or an insulation fault location system may be provided to indicate an audible and/or visual signal, which shall continue as long as the fault persists.

If there are both audible and visual signals it is permissible for the audible signal to be cancelled, but the visual alarm shall continue as long as the fault persists.

*Note: It is recommended that a first fault be eliminated with the shortest practical delay.*

After the occurrence of a first fault, conditions for automatic disconnection of supply in the event of a second fault occurring on a different live conductor shall be as follows:

**a)** Where exposed-conductive-parts are interconnected by a protective conductor collectively earthed to the same earthing system, the conditions similar to a TN system apply and the following conditions shall be fulfilled where the neutral conductor is not distributed in a.c. systems and in d.c. systems where the mid-point conductor is not distributed:

$2 I_a Z_s \leq U$

or where the neutral conductor or mid-point conductor respectively is distributed:

$2 I_a Z'_s \leq U$

where

- $U_0$ is the nominal a.c. or d.c. voltage in V, between line conductor and neutral conductor or mid-point conductor, as appropriate
- $U$ is the nominal a.c. or d.c. voltage in V between line conductors
- $Z_s$ is the impedance in Ω of the fault loop comprising the line conductor and the protective conductor of the circuit
- $Z'_s$ is the impedance in Ω of the fault loop comprising the neutral conductor and the protective conductor of the circuit
- $I_a$ is the current in A causing operation of the protective device within the the required time for TN systems

b) Where the exposed-conductive-parts are earthed in groups or individually, the following condition applies:

$R_A \times I_a \leq 50$ V

where

$R_A$ is the sum of the resistance in $\Omega$ of the earth electrode and the protective conductor for the exposed-conductive-parts

$I_a$ is the current causing automatic disconnection of the disconnection device in a time complying to that for TT systems

## 4.6 Additional Protective Measures

Beside the standard distribution systems TN, TT and IT systems, IEC 60364-4-41 gives information on several other protective measures.

### 4.6.1 Functional Extra-Low Voltage (FELV)

Where, for functional reasons, a nominal voltage not exceeding 50 V a.c. or 120 V d.c. is used but all the requirements of extra-low voltage relating to SELV or to PELV are not fulfilled, and where SELV or PELV is not necessary, the supplementary provisions shall be taken to ensure basic protection and fault protection. This combination of provisions is known as FELV.

The sources for FELV system shall be either a transformer with at least simple separation between windings or shall have electrical separation.

Plugs and socket-outlets for FELV systems shall comply with the following requirements:

- plugs shall not be able to enter socket-outlets of other voltage systems
- socket-outlets shall not admit plugs of other voltage systems, and
- socket-outlets shall have a protective conductor contact

## 4.7 Protective Measure: Double or Reinforced Insulation

In general double or reinforced insulation is a protective measure in which
- basic protection is provided by basic insulation, and fault protection is provided by supplementary insulation, or
- basic and fault protection is provided by reinforced insulation between live parts and accessible parts

Where this protective measure is to be used as the sole protective measure (i.e. where a whole installation or circuit is intended to consist entirely of equipment with double insulation or reinforced insulation), it shall be verified that the installation or circuit concerned will be under effective supervision in normal use so that no change is made that would impair the effectiveness of the protective measure. This protective measure shall not therefore be applied to any circuit that includes a socket-outlet or where a user may change items of equipment without authorization.

### 4.7.1  Requirements for Basic Protection and Fault Protection

#### 4.7.1.1  Electrical Equipment

Where the protective measure of double or reinforced insulation, is used for the complete installation or part of the installation, electrical equipment shall be of one of the following types, and type tested and marked to the relevant standards:

- electrical equipment having double or reinforced insulation (Class II equipment)
- electrical equipment declared in the relevant product standard as equivalent to Class II, such as assemblies of electrical equipment having total insulation
- electrical equipment having basic insulation only shall have supplementary insulation applied in the process of erecting the electrical installation, providing a degree of safety equivalent to electrical equipment with enclosures
- electrical equipment having uninsulated live parts shall have reinforced insulation applied in the process of erecting the electrical installation, providing a degree of safety equivalent to electrical equipment (see below) and complying with the protective requirements for enclosures; such insulation being recognized only where constructional features prevent the application of double insulation

#### 4.7.1.2  Enclosures

In general electrical equipment being ready for operation, all conductive parts separated from live parts by basic insulation only, shall be contained in an insulating enclosure affording at least the degree of protection IPXXB or IP2X.

There are several requirements for enclosures described in IEC 60364-4-41, 412.2.2.

#### 4.7.1.3  Installation

The installation of equipment, for example, fixing, connection of conductors, etc., (see 4.7.1.1) shall be effected in such a way as not to impair the protection afforded in compliance with the equipment specification.

#### 4.7.1.4 Wiring Systems

Wiring systems installed in accordance with IEC 60364-5-52, Selection and erection of electrical equipment – wiring systems, are considered to meet the requirements for basic and fault protection, if:
- the rated voltage of the wiring system shall be not less than the nominal voltage of the system and at least 300/500 V, and
- adequate mechanical protection of the basic insulation

## 4.8 Protective Measure: Electrical Separation

Electrical separation is a protective measure in which
- basic protection is provided by basic insulation of live parts or by barriers and enclosures in accordance with Annex A of IEC 60364-4-41, and
- fault protection is provided by simple separation of the separated circuit from other circuits and from earth.

## 4.9 Protective Measure: Extra-Low-Voltage Provided by SELV and PELV

In general protection by extra-low-voltage is a protective measure which consists of either two different extra-low-voltage systems:
- SELV (Safety Extra-Low Voltage); or
- PELV (Protective Extra-Low Voltage)

This protective measure requires:
- limitation of voltage in the SELV or PELV system to the upper limit of voltage Band I, 50 V a.c. or 120 V d.c., and
- protective separation of the SELV or PELV system from all circuits other than SELV and PELV circuits, and basic insulation between the SELV or PELV system and other SELV or PELV systems, and
- for SELV systems only, basic insulation between the SELV system and earth

## 4.10 Additional Protection

Additional protection may be specified with the protective measure under certain conditions of external influence and in certain special locations (see the corresponding Part 7 of IEC 60364)[1].

---

[1] See Chapter 17

### 4.10.1 Additional Protection: Residual Current Protective Devices (RCDs)

The use of residual current protective devices (RCDs) with a rated residual operating current not exceeding 30 mA, is recognized in a.c. systems as additional protection in the event of failure of the provision for basic protection and/or the provision for fault protection or carelessness by users.

The use of such devices is not recognized as a sole means of protection and does not obviate the need to apply one of the protective measures, described in the previous clauses.

### 4.10.2 Additional Protection: Supplementary Equipotential Bonding

- Supplementary protective equipotential bonding is considered as an addition to fault protection.
- The use of supplementary protective bonding does not exclude the need to disconnect the supply for other reasons, for example protection against fire, thermal stresses in equipment, etc.
- Supplementary protective bonding may involve the entire installation, a part of the installation, an item of apparatus, or a location.
- Additional requirements may be necessary for special locations, (see the corresponding Part 7 of IEC 60364), or for other reasons.

Supplementary protective equipotential bonding shall include all simultaneously accessible exposed-conductive-parts of fixed equipment and extraneous-conductive-parts including where practicable the main metallic reinforcement of constructional reinforced concrete. The equipotential bonding system shall be connected to the protective conductors of all equipment including those of socket-outlets.

Where doubt exists regarding the effectiveness of supplementary protective equipotential bonding, it shall be confirmed that the resistance $R$ between simultaneously accessible exposed-conductive-parts and extraneous-conductive-parts fulfils the following condition:

$$R \leq \frac{50 \text{ V}}{I_a} \quad \text{in a.c. systems}$$

$$R \leq \frac{120 \text{ V}}{I_a} \quad \text{in d.c. systems}$$

where

$I_a$ is the operating current in A of the protective device
- for residual current protective devices (RCDs), $I_{\Delta n}$
- for overcurrent devices, the 5 s operating current

## 4.11 Appendices of IEC 60364-4-41

There are three appendices to the standard IEC 60364-4-41:

- **Annex A (normative): Provisions for basic protection**

  Provisions for basic protection provide protection under normal conditions and are applied where specified as a part of the selected protective measure.
    - basic insulation of live parts – intended to prevent contact with live parts, where live parts are completely covered with insulation, removable only by destruction
    - barriers and enclosures – intended to prevent contact with live parts, with suitable and practicable precautions against unintentional contact with live parts

- **Annex B (normative): Obstacles and placing out of reach**

  The protective measure of obstacles and placing out of reach are for application in installations with or without fault protection that are controlled or supervised by skilled or instructed persons, by
    - Obstacles shall prevent unintentional contact with live parts, but not intentional contact by deliberate circumvention of the obstacle.
    - Placing out of reach – only to prevent unintentional contact with live parts.

- **Annex C (normative): Protective measures for application only when the installation is controlled or under supervision of skilled or instructed persons for:**
    - Non-conducting location
    - Protection by earth-free local equipotential bonding
    - Electrical separation for the supply of more than one item of current-using equipment

# 5 Unearthed IT Systems

The principle of unearthed IT systems with insulation monitoring has been in place in IEC standards since 1983. It is described in the International standards IEC 60364-1 and IEC 60364-4-41 (see Chapter 4).

The IT system is usually equipped with equipotential bonding and insulation monitoring devices. IT systems are supplied by a transformer, generator, battery or by an independent voltage source. The distinctive feature of these a.c. and d.c. systems is the fact that no live conductor of the system is directly earthed. This has the advantage that first earth faults or short-circuits to exposed conductive parts do not interfere with the electrical equipment. In the event of a short-circuit to exposed conductive parts in a three-phase system, the faulty conductor takes over the protective conductor potential; the two other conductors are increased to the phase-to-phase voltage. The sum of the leakage currents of the fault-free phase conductors flow as a capacitive fault current. These leakage currents must be kept low enough, to prevent unintentional contact with live parts. If the fault current in the event of a short-circuit to exposed conductive parts is low, disconnection of the equipment is unnecessary. However, measures must be taken to avoid the occurance of a second fault occurs.

In the IT system the exposed conductive parts of the equipment are being connected to the protective conductor (supplementary equipotential bonding) in such a way that the upstream fuse is triggered in the case of a second fault at another conductor. However, in order to already receive a warning at the first insulation fault, monitoring of the insulation level of the installation is required. Thus the fault in the equipment may be located and monitored without excitement. For continuous monitoring of the insulation status of the IT system, insulation monitoring devices are installed between phase conductors and protective conductors which continuously measure the insulation resistance and give visual or acoustic signals, if the insulation resistance falls below a minimum value.

For the safety of the operation it is highly recommended to detect and correct an insulation fault as soon as possible. Portable or stationary insulation fault location systems are available for that purpose. In accordance with IEC 60364-4-41 any type of protective device is allowed in the system, however, the use of insulation monitoring devices is the preferred method for the IT system.

Ultimately, the special advantages of the IT system in regards to operating, fire and shock protection may only be of use with suitable insulation monitoring systems.

## 5.1 Example of IT Systems with Equipotential Bonding and Insulation Monitoring

Inadequate insulation of an electrical device or an electrical installation is the most common cause for dangerous touch voltages or electrical fires. To prevent such hazards, it is appropriate to permanently monitor the insulation level in electrical installations and to send a warning signal as soon as a critical value is reached. Only then is it possible to take the necessary preventive measures before a critical insulation fault or break of the power supply can occur. One such protective measure is the IT system with equipotential bonding and insulation monitoring.

As already mentioned, the IT system is supplied by a transformer or by an independent voltage source. The special characteristic of this kind of system is that no active conductor is directly connected to earth.

In the most common IT system, the configuration of the exposed conductive parts of all connected consumers, all accessible conductive building structures, pipework, lightning protection, and other earthed devices connected to the protective conductor (**Figure 5.1**) are called supplementary protective equipotential bonding.

A reduction of the value of the PE resistance by interconnection of all conductive, earthed parts, brings a reduction of the touch voltage. A reduction of the fault current may also bring a reduction of the touch voltage. For this reason, the line conductors of the IT system must not be connected to the protective conductor at any point. In IT systems, in the event of a low resistance fault to exposed conductive parts or to earth, only a low capacitive current will flow. Owing to the lack of a return path, only a low fault current will flow dependent on the value of the insulation resistances $R_F$ and the capacitance $C_e$ of the conductors to earth (**Figure 5.2**).

In d.c. IT systems dependency on $C_e$ does not exist. **Figure 5.3** shows the construction of such a system.

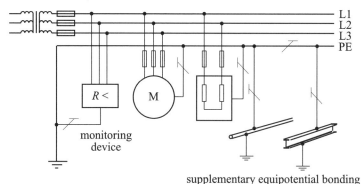

**Figure 5.1** Design of an IT system with supplementary protective equipotential bonding and insulation monitoring

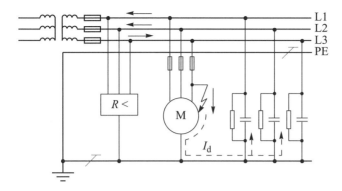

**Figure 5.2** Path of the fault current $I_d$ in the event of a short circuit to exposed conductive parts of a three-phase IT system

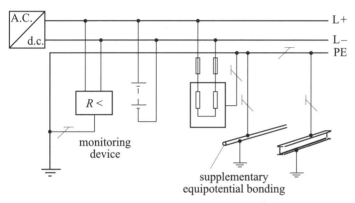

**Figure 5.3** Design of a d.c. IT system with supplementary protective equipotential bonding with insulation monitoring

The difference between an earthed system and an unearthed system in the event of a fault is apparent by a comparison of **Figure 5.4** and **Figure 5.5**. On the occurrence of an insulation fault $R_F$ in an earthed system an earth-fault-current $I_d$ is flowing which is equal to the short-circuit current $I_k$. The upstream fuse is triggered, and the power supply is interrupted (Figure 5.4).

The unearthed system reacts differently (Figure 5.5). The diagram shows that in the event of an insulation fault $0 \leq R_F \leq \infty$ no more than a small capacitive current flows through the line capacitances. The upstream fuse is not triggered and the power supply is not interrupted (Figure 5.5).

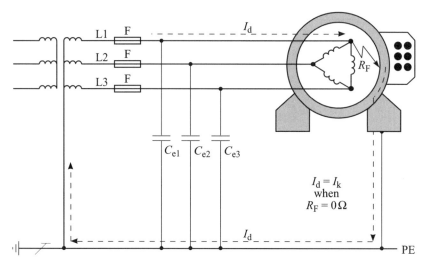

**Figure 5.4** Insulation fault in a TN system

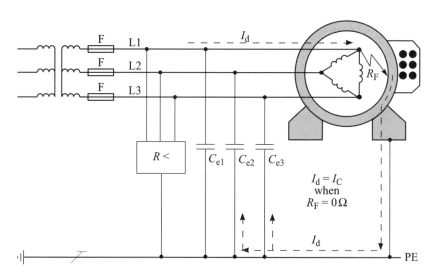

**Figure 5.5** Insulation fault in an IT system

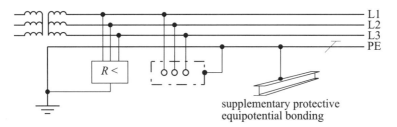

**Figure 5.6** IT system with supplementary protective equipotential bonding

## 5.2 Supplementary Protective Equipotential Bonding in the IT System

In the IT system supplementary protective equipotential bonding has to be provided, if the conditions for tripping for overcurrent protective devices are not met.

The underlying idea for the protection against indirect touch by supplementary protective equipotential bonding can be outlined as follows:

Each exposed conductive part is connected to other exposed and extraneous conductive parts via the supplementary protective equipotential bonding, which are within the limits of arm's reach. Cross-section and conductor routing shall be selected in a way, that, in the case of a short circuit to exposed conductive part, it is impossible to bridge dangerous touch voltages within the limits of arm's reach.

The supplementary protective equipotential bonding shall include all simultaneous exposed conductive parts of permanent operation equipment and all simultaneous extraneous conductive parts (**Figure 5.6**). The equipotential bonding system shall be connected to the protective conductors of all operating equipment, including socket-outlets.

Supplementary protective equipotential bonding shall be implemented with equipotential bonding conductor according to IEC 60364-5-54.

### 5.2.1 Minimum Cross-Sections for Supplementary Protective Equipotential Bonding

In the IT system the supplementary protective equipotential bonding corresponds with the protective conductor. It is recommended to select the respective cross-sections according to **Table 5.1**. In the IT system the protective conductor always runs separately to the other conductors, hence the cross-section may be reduced.

The requirements for the cross-sections of the supplementary protective equipotential bonding are stated in **Table 5.1**.

| Cross-section for equipotential bonding conductors of the supplementary protective equipotential bonding | | |
|---|---|---|
| Normal | Between two parts | 1x cross-section of the smaller protective conductor |
| | Between one part and another exposed conductive part | 0,5 x cross section of the protective conductor |
| Minimum | At mechanical protection | 2,5 mm² Cu or Al*) |
| | Without mechanical protection | 4 mm² Cu or Al*) |

*) If conductors made of aluminium (Al) are run unprotected, the danger of conductor disconnection because of possible corrosion and low mechanical strength exists.
Cu= Copper, Al= Aluminium

**Table 5.1** Cross-section of the supplementary protective equipotential bonding

## 5.3 Testing of IT Systems According to IEC 60364-6

In August 2001 the second edition of the International standard IEC 60364-6-61 was published under the title: Electrical installations of buildings – Part 6-61: Verification – Initial verification, replacing the 1986 edition and two amendments (1993 and 1997). This edition is now likewise superseded by a completely revised standard. This constitutes a technical revision and includes all parts of chapter 6. The new IEC 60364-6[1] is titled "Low-voltage electrical installations – Part 6: Verification". It provides requirements for initial and periodic verification of an electrical installation. Clause 61 gives the information on initial verification.

In general every installation shall now be verified during erection, as reasonably practical, and on completion, before commissioning by the user.

The initial verification shall be performed by a skilled person, competent in verification. Inspection shall precede testing and shall normally be done prior to energizing the installation.

### 5.3.1 Testing

The test methods described in this clause are given as reference methods; other methods are not precluded, provided they give no less valid results.

Measuring instruments and monitoring equipment and methods shall be chosen in accordance with the relevant parts of IEC 61557. If other measuring equipment is used, it shall provide no less degree of performance and safety.

---

[1] The author points out that this new 3rd edition of IEC 60364-6 makes references to IEC 60364-4-41, Ed. 4.0:2001 – and not to Ed. 5.0:2005, as described in chapter 4!

The following tests shall be carried out where relevant and should preferably be made in the following sequence:

a) continuity of conductors

b) insulation resistance of the electrical installation

c) protection by SELV, PELV or by electrical separation

d) floor and wall resistance/impedance

e) automatic disconnection of supply

f) additional protection

g) polarity test

h) test of the order of the phases

i) functional and operational tests

j) voltage drop

In the event of any test indicating failure to comply, that test and any preceding test, the results of which may have been influenced by the fault indicated, shall be repeated after the fault has been rectified.

**5.3.2  Verification for IT Systems**

In general the verification of the effectiveness of the measures for protection against indirect contact by automatic disconnection of supply for IT systems is effected as follows:

Compliance with the rules of 413.1.5.3 of IEC 60364-4-41:2001-08 shall be verified by calculation or measurement of the current $I_d$ in case of a first fault at the line conductor or at the neutral.

*Note: The measurement is made only if the calculation is not possible, because all the parameters are not known. Precautions are to be taken while making the measurement in order to avoid the danger due to a double fault.*

Where conditions that are similar to conditions of TT systems occur, in the event of a second fault in another circuit (see point a) of 413.1.5.5 of IEC 60364-4-41:2001-08), verification is made as for TT systems (see point b) of this clause).

Where conditions that are similar to conditions of TN systems occur, in the event of a second fault in another circuit (see point b) of 413.1.5.5 of IEC 60364-4-41: 2001-08), verification is made as for TN systems (see point a) of this clause).

## 5.4 Protection Against Overcurrent in All Distribution Systems

For protection against overcurrent in IT systems, the IEC standard IEC 60364-4-43 provides a helpful tool. Part 4-43[2] of this series of standards, subtitled "Protection for safety – Protection against overcurrent", provides requirements for the protection of live conductors from the effect of overcurrent. It describes how live conductors are protected by one or more devices for automatic interruption of the supply in the event of overload and short-circuits. Coordination of overload protection and short-circuit protection is also covered.

A general requirement is that protective devices are provided to break any overcurrent in the circuit conductors before such a current could cause a danger due to thermal or mechanical effects detrimental to insulation, joints, terminations or surroundings of the conductor.

### 5.4.1 Protection of Line Conductors

Detection of overcurrent shall be provided for all line conductors, with a few exceptions for TN, TT and IT systems

Usually all line conductors in IT systems are equipped with a protective device and monitored with an insulation monitoring device.

In IT systems the overcurrent can only flow, if two faults happen at the same time on two different live conductors.

### 5.4.2 Protection of the Neutral Conductor

#### 5.4.2.1 TT or TN Systems

Where the cross-sectional area of the neutral conductor is at least equivalent to that of the line conductors, it is not necessary to provide overcurrent detection for the neutral conductor or a disconnecting device for that conductor.

When the cross-sectional area of the neutral conductor is less than that of the line conductors, it is necessary to provide overcurrent detection for the neutral conductor, appropriate to the cross-sectional area of that conductor; this detection shall cause the disconnection of the phase conductors, but not necessarily of the neutral conductor.

In every case the neutral conductor shall be protected against a short circuit.

---

[2] The author points out that IEC 60364-4-43, Ed. 3.0: Low-voltage electrical installations – Part 4-43: Protection for safety – Protection against overcurrent: Ed. 2.0:2001 is currently under review and expected to be published in 2007, yet the contents of this standard even at this development stage is important enough to be included in this chapter (see Chapter 1, Table 1.2).

## 5.4.3 IT Systems

Where the neutral conductor is distributed, it is necessary to provide overcurrent detection for the neutral conductor of every circuit. The overcurrent detection shall cause the disconnection of all the live conductors of the corresponding circuit, including the neutral conductor. This measure is not necessary if

- the particular neutral conductor is effectively protected against short-circuit by a protective device placed on the supply side, for example at the origin of the installation; or if
- the particular circuit is protected by a residual current-operated protective device with a rated residual current not exceeding 0.20 times the current-carrying capacity of the corresponding neutral conductor. This device shall disconnect all the live conductors of the corresponding circuit, including the neutral conductor. The device shall have sufficient breaking capacity for all poles.

## 5.4.4 Protection Against Overload

### 5.4.4.1 Co-Ordination Between Conductors and Overload Protective Devices

The operating characteristics of a device protecting a cable against overload shall satisfy the two following conditions:

$$I_B \leq I_n \leq I_Z \tag{5.1}$$

$$I_2 \leq 1{,}45 \times I_Z \tag{5.2}$$

where

$I_B$ is the design current for that circuit

$I_Z$ is the continuous current-carrying capacity of the cable

$I_n$ is the nominal current of the protective device

*Note: For adjustable protective devices, the nominal current $I_n$ is the current setting selected.*

$I_2$ is the current ensuring effective operation in the conventional time of the protective device

The current $I_2$ ensuring effective operation of the protective device shall be provided by the manufacturer, or as given in the product standard.

Protection in accordance with this clause may not ensure protection in certain cases, for example where sustained overcurrents less than $I_2$ occur. In such cases, consideration should be given to selecting a cable with a larger cross-sectional area.

### 5.4.4.2 Omission of Devices for Protection Against Overload

The various cases stated in this subclause shall not be applied to installations situated in locations presenting a fire risk or risk of explosion and where the requirements for special installations and locations specify different conditions.

In general devices for protection against overload need not be provided for

a) a conductor situated on the load side of a change in cross-sectional area, nature, method of installation or in constitution, that is effectively protected against overload by a protective device placed on the supply side,

b) a conductor that is not likely to carry overload current, provided that this conductor is protected against short-circuit and that it has neither branch circuits nor socket-outlets,

c) at the origin of an installation where the distributor provides an overload device and agrees that it affords protection to the part of the installation between the origin and the main distribution point of the installatoin where further overload is provided.

### 5.4.4.3 Position or Omission of Devices for Protection Against Overload in IT Systems

The provisions for an alternative position or omission of devices for protection against overload are not applicable to IT systems unless each circuit not protected against overload is protected by one of the following means:

a) use of the protective measures described in 413.2 of IEC 60364-4-41:2001-08,

b) protection of each circuit by a residual current protective device (RCD) that will operate immediately on the second fault,

c) for permanently supervised systems only use of insulation monitoring which either

 – causes the disconnection of the circuit when the first fault occurs, or

 – gives a signal indicating the presence of a fault. The fault shall be rectified according to the operational requirements and recognizing the risk from a second fault

*Note: It is recommended to install an insulation fault location system according to IEC 61557-9. With the application of such a system it is possible to detect and locate the insulation fault without interruption of the supply.*

In IT systems without a neutral conductor the overload protective device may be omitted in one of the phase conductors if a residual current protective device (RCD) is installed in each circuit.

### 5.4.5 Cases where Omission of Devices for Overload Protection shall be Considered for Safety Reasons

The omission of devices for protection against overload is permitted for circuits supplying current-using equipment where unexpected disconnection of the circuit could cause danger or damage.

Examples of such cases are:
- exciter circuits of rotating machines
- supply circuits of lifting magnets
- secondary circuits of current transformers
- circuits which supply fire extinguishing devices
- circuits supplying safety services (burglar alarm, gas alarms, etc)

*Note: In such cases consideration should be given to the provision of an overload alarm.*

## 5.5 Connection of Insulation Monitoring Devices (IMDs)

According to the International standard IEC 61010-1, „Safety requirements for electrical equipment for measurement, control and laboratory use – Part 1: General requirements", the electrical installation should be protected against the spread of fire outside the equipment in normal condition or in single fault condition. This also applies to insulation monitoring devices in IT systems.

The risk of ignition and occurrence of fire may be reduced to a tolerable level by selecting the proper material during the design of the devices, for example by using flame retardant or self-extinguishing materials.

The risk of an electric arc as a reaction to a short-circuit in the device may be problematical and may only be prevented by special constructive measures, for example by mechanical fixation of printed circuit-boards and components, enclosures, etc., or by an appropriate protection against short-circuit. IEC 61010-1 does not specifically regulate the provision of an overcurrent protective device, but if one is fitted, the manufacturer's instructions shall specify the overcurrent protection devices required in the building installation. If an overcurrent protection is provided, it shall be within the equipment.

### 5.5.1 Coupling and Fuse Protection

If insulation monitoring devices are connected to the circuit, short-circuit protection may be omitted, if the wiring system is in such a way, that the risk of short-circuit is reduced to a minimum. If the wiring system is short-circuit and earth-fault proof, the risk of short-circuit is reduced to the constructive design of the device.

Important aspects for risk assessment are:
- Type of internal resistance (inductive, resistive)
- Specifications of the terminals
- Mechanical specifications
- Existing suppressor circuits

Connected insulation monitoring devices without short-circuit protection give the advantage of "uninterruptible monitoring", which means that the indicated insulation resistance by the insulation monitoring device is not falsified because of a defect coupling fuse. In this way, forgoing protection has certain cost advantages.

If coupling includes short-circuit protection, it is recommended to apply insulation monitoring devices with "coupling-monitoring", which means that this type includes short-circuit protection with monitoring the coupling process. In this way a high level of safety is offered in combination with suitable insulation monitoring. **Figure 5.7** gives an example of an insulation monitoring device with permanent monitoring of the coupling process and protective conductor connection, also including indication in the event of circuit interruption.

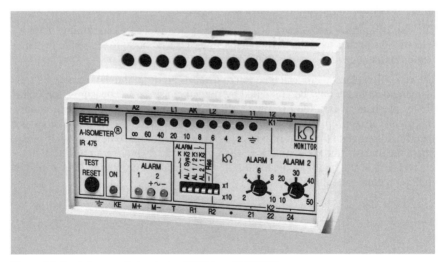

**Figure 5.7** Insulation monitoring device (IMD) type IR475LYX
[Photo supplied by Bender, Gruenberg, Germany]

## 5.5.2 Auxiliary Supply and Fuse Protection

The connection of the supply voltage of insulation monitoring devices shall be provided with a protective device against short-circuits. If no internal protection against short-circuit is included in the device, a protective device on the supply side shall be applied. Documentation by the manufacturer usually includes respective instructions. It is common practice that a 6 A fuse is used.

# 6 Special Features and Advantages of IT Systems

Electrical distribution systems with an earthed neutral point are generally well known. IT systems are completely insulated from earth, but are still fairly unknown. Every unearthed system is a separate system with its own transformer and its own power generator or battery. They are applied in all areas where a high degree of operating safety, fire safety and accident prevention is required.

In many countries these IT systems are therefore mandatory, recommended or used for other reasons in following applications:
- Low-voltage electrical installations[1] with nominal voltages up to 1000 V
- Operating theatres, anaesthetic rooms and intensive care units in hospitals
- Emergency lighting in communal facilities
- Open cast and underground mining
- Ships
- Control and regulating circuits
- Furnaces
- Steelworks
- Electrical power plants
- Chemical industry
- Operations with explosive environment
- Highly sensitive production processes
- Test and laboratory installations
- Electrical supply systems for electronic hardware
- Electrical installations in railway systems
- Floor and ceiling heating systems
- Information technology systems
- Electrical equipment of machines
- Electro vehicles
- Telecommunications
- Stand-by generators
- Electrical diving equipment
- Building sites

---

[1] New terminology according to IEC 60364-4-41:2005-12

Because of this wide field of application, it is of paramount importance that respective users get familiar with the special characteristics of IT systems.

Compared to solidly earthed systems, IT systems have the following advantages:
- higher operational safety
- higher fire safety
- better accident prevention due to limited touch currents
- higher permissible earthing resistance
- information advance

Certainly these advantages may not be available in every application, but there are still highly convincing reasons to prefer the IT system.

The disadvantage, on the other hand is, that the insulation fault is not automatically indicated by a blown fuse. A certain amount of training and a sound knowledge of the operational conditions are needed for fault location, particularly in large installations.

Insulation fault location systems in accordance with IEC 61557-9 are a great reference to the electrical engineering technicians.

## 6.1 Higher Operating Safety

The following are a few examples of the advantages of IT systems in regards to operational reliability:
- insulation monitoring for higher reliability of the system, is only possible in IT systems
- a conductor may be completely short-circuited to earth without interruption to operation
- preventive maintenance by continuous monitoring and indication of the insulation state of the installation is possible
- early detection of defective equipment through immediate indication when devices are connected
- monitoring of de-energized systems
- monitoring of d. c. systems
- normal operation of control systems even in the event of partial or complete insulation faults

This applies in particular to auxiliary circuits:

Insulation faults rarely occur in adequately maintained auxiliary circuits, since weak points in the insulation system can be detected and eliminated at an early stage. They are most unlikely to occur in proper control cabinets. If insulation faults occur in external lines, spontaneous low-resistance earth faults may be expected to occur in normal and dry operating conditions, rather than in gradually deteriorating insulation, bordering on the response thresholds of the insulation monitoring device.

In electrical installations with a reduction of dielectric strength, for example through humidity or conductive dust and, provided contact is sufficient, lower nominal voltages shall be applied, since they reduce the danger by creepage currents.

For extensive control systems a subdivision into two or more independent auxiliary circuits should be considered, since this will further reduce the fault probability and facilitate fault location for the operating staff.

## 6.2 Improving Fire Safety

For improving fire safety, the unearthed IT system has the following advantages:
- gradual deterioration of the insulation resistance can be detected and remedied immediately
- the occurrence of accidental arcs, which are common causes of fires, is prevented (except for extremely high-output systems)
- equipment or parts of equipment exposed to fire or explosion hazards may be separated from the rest of the system by means of isolating transformers and may be operated as a separate small system, retaining all the advantages of super-sensitive automatic monitoring
- valuable equipment, for example motors, cannot be damaged by electric arcs in the event of an incomplete short circuit to exposed conductive parts

**Figure 6.1** shows how the point of the insulation fault may be heated through an earth fault, since in the earthed system, the circuit is being closed by the neutral point earth connection of the isolating transformer.

Temperature rise at the fault location is determined by energy loss in the fault resistance, which likewise determines the level of the fire risk. However, a fire can only break out, if the following conditions are fulfilled [6.1]:

- combustible or flammable material
- oxygen
- right quantitative ratio
- ignition temperature

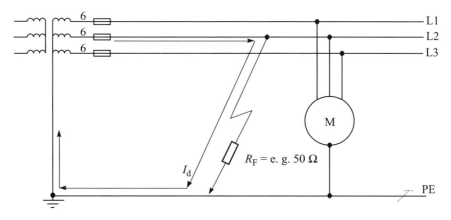

**Figure 6.1** Diagram of an earthed TN system

The ignition temperature is the only element in direct correlation to the electrical current. Whether the ignition temperature is reached does not only depend on the energy loss in the fault resistance, but also on the thermal-specific characteristics of the fault location.

The energy of heat $H$ is created by a current $I_d$ on a fault $R_F$ for a certain duration $t$, or:

$$H = I_d^2 \cdot R_F \cdot t \qquad (6.1)$$

where

$H$ electrical energy
$R_F$ insulation fault
$I_d$ fault current in A of the first fault
$t$ duration of the fault

Since $R_F$ may be anywhere between "infinite" and "zero", it is impossible to pre-determine the dissipated energy. The wide-spread opinion, that the short-circuit is the cause of a fire, can be laid to rest. As proofed by the equation (6.1), $H = 0$, when $R_F = 0$ kΩ. However it is rare that the 0 Ω fault occurs in real life. It is rather more likely that in the earthed system a fault resistance will occur at the fault location, which, under heat, develops into a short circuit and disconnection of the system. **Figure 6.2** shows the described principle with the same type of fault constellation, but as an unearthed IT system. Here the thermal energy generated at the fault resistance is low. The flowing current is only determined by the size of the leakage capacitances, but these are often negligible. Therefore in unearthed IT systems, the minimum values needed to create a fire at the fault resistance with an energy loss of 60 W, a current of at least 0,3 A and more than 5 W, are normally not reached.

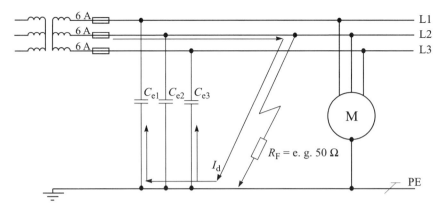

**Figure 6.2** Diagram of an unearthed IT system

## 6.3 Improving Accident Prevention with Limited Touch Voltages

IT systems play a special role in regards to protecting humans from the dangers of electric shock, because:
- In small and medium-sized installations the earth leakage currents and hence the highest possible touch voltages are kept so low that persons are not injured if direct contact occurs between conductor and earth.

As already shown in previous figures, the maximum fault currents in IT systems are determined only by the internal resistance of the insulation monitoring devices and by the existing system leakage capacitances. The maximum currents that may pass through the human body are usually below the hazard threshold.

If there is good protective earthing and low earthing resistance of the system earthing, in case of a first fault, then the expected touch voltage in IT systems is very low.

However, it is not meant to create the impression that IT systems provide protection against electric shock at direct contact. The possible shock current in the a.c. IT system is determined primarily by the system leakage capacitances and the body resistance. Since large industrial systems are often operated as unearthed systems, it is only natural that high shock currents may flow.

## 6.4 Higher Permissible Earthing Resistance

Another important advantage of IT systems, as opposed to TN systems, is the higher permissible earthing resistance. This advantage is frequently used in the application of portable generators (**Figure 6.2**) by fire brigades, relief organisations, the Red Cross or Armed Forces [6.2] in order to provide sufficient protection in emergencies, in the face of unknown earthing conditions.

According to IEC 60364-4-41:

"Exposed conductive parts shall be earthed individually, in groups or collectively. The following condition shall be fulfilled:

$$R_A \times I_d \leq 50\,V \tag{6.2}$$

where

$R_A$ is the resistance in $\Omega$ of the earth electrode and protective conductor for the exposed-conductive-parts

$I_d$ is the fault current in A of the first fault of negligible impedance between a line conductor and an exposed-conductive-part; the value of $I_d$ takes account of leakage currents and the total earthing impedance of the electrical installation." (IEC 60364-4-41, 411.6.2)

IEC 60364-5-55:2002-06, "Electrical installations of buildings – Part 5-55: Selection and erection of electrical equipment – Other equipment", clause 551.6, "Standby systems", takes this into consideration. It says, that protective measures against

**Figure 6.2** Portable Generator with insulation monitoring device (Photo supplied by Honda, Germany)

electric shock in case of a fault without automatic disconnection at the first fault are preferred. In IT systems, continuous insulation monitoring devices shall be provided which give an audible or visual indication of a first fault.

## 6.5  Information Advantage with IT Systems

The previous chapters dealt with the different types of distribution systems and possible protective measures as defined in current standards. However, the aspects of effectiveness of protective measures, preventive maintenance, availability of the installation, cost analysis or maintenance expenditure remained unconsidered.

Some of the tangible and verifiable advantages of the supply with IT systems have been pointed out. Let us now take a closer look at the practical implications of terminology in regards to the advantages of IT systems:

- higher operating safety
- improved fire safety
- improved accident prevention
- higher permissible earthing resistance

How is this information put into practice? How is this expressed in monetary value? It is not enough to install supply systems as IT systems and then to lean back in anticipation of all the advantages of such systems. Careful planning and the proper design of the IT system is important for guaranteeing the safety aspects. Insulation monitoring devices are an integral part of IT systems, as availability rises and falls with their proper selection. The advantage of high availability of IT systems all depends on the right selection of insulation monitoring devices. Wrong planning only leads to seeming safety. Insulation monitoring devices which give wrong or faulty measures may negate the advantages of IT systems. This has to be considered during the planning phase of an electrical installation.

In principle and independently from the type of distribution systems the following applies: the determining factor for the availability of an electrical installation is the insulation resistance. It is for good reason the priority on the list of electrical safety measures.

The causes for the deterioration of the insulation are manifold and often very common: humidity, moisture, deterioration, contamination, environmental and climatic influences. In addition unpredictable circumstances may happen, such as the shovel of an excavator accidentally damaging a supply cable or the drill cutting the cable inside the wall. The list of possible effects of insulation faults is long and the effects are ranging from being inconvenient to dangerous or fatal:

- sudden shutdown of the installation, disruption of important production processes or procedures

- maloperation due to several insulation faults occurring at the same time. If two insulation faults coincide, the effect might simulate that of a control switch
- fire danger because of failure through high-resistance insulation faults. Just 60 W at the fault location is regarded as a high fire-risk level
- break-downs or interruptions of critical equipment (for instance in the healthcare environment or in aviation)
- tedious, difficult and therefore expensive search for the fault location

Even a brief interruption to the electrical supply may trigger effects resulting in costly downtime, the extend is often difficult to assess. The massive public power supply failure that hit the east coast of the US in 2003 are an evident example not of just financial damages, but also of the dangers to human lives, to sensitive components and malfunctions in critical equipment. The financial damage caused by a breakdown of the lighting installations in a museum, for instance is different from that of a failure of stirring equipment being operated in a chemical plant. The cause of the failure may be the same in both cases: perhaps a slowly developing and progressing insulation fault, which can suddenly cause damage or trigger an upstream fuse.

It is not surprising that fault analysis reports from corporations or electric commissions for example list insulation and earth faults as the number one cause for operational breakdowns (also refer to chapter 7 on insulation resistance). Practical experience has shown that insulation faults develop gradually and rarely occur suddenly.

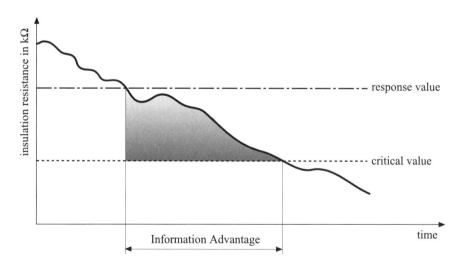

**Figure 6.3** Advanced information by monitoring the insulation resistance

Residual current protective devices (RCDs) and circuit-breaker, which have been applied as part of the protective measures for a long time now, however only react to those sudden insulation faults. No measures have yet been offered to protect from and monitor the slowly developing insulation faults.

It is important now to do something against the effects of those gradually developing insulation faults. With an intelligent insulation monitoring system, suitable for the electrical installations, it is possible to detect any drop in the insulation resistance in time. This in-time information provides the operator with an advantage which can be utilized for his corrective measures and preventive maintenance (see **Figure 6.3**).

This information advantage and hence the high availability of the installation provide economic efficiency. But it has to be properly planned and measures taken to realize this goal.

The type of measures needed to be taken depend on various factors: the importance and quality of the supply, the frequency of application, the types of consumers, the size of distribution system, the maintenance frequency are only some.

It is often enough to apply a properly selected, adequately planned insulation monitoring device in an IT system. On the other hand in other installations more extensive monitoring is called for perhaps with a complete insulation-fault-location system. It is up to the designers and project managers of electrical installations to choose the right insulation monitoring device for the specific IT system. Existing installations may require the scrutinized eye of the operator and possibly the adaptation of the insulation monitoring device to the existing conditions and specific requirements.

It brings tangible advantages in terms of operating availability and safety of the installation to the operator, if he knows how to use the information advantage IT systems offer. This is unique to IT systems no other distribution systems provide these.

### 6.5.1 Maintenance of Electrical Supply Systems

The question of availability of electrical installations and preventive maintenance are directly or indirectly addressed in various standards. In order to prevent the costly effects of a broken down electrical supply, an appropriate maintenance procedure is needed. This includes regular tests, such as testing the insulation resistance. The purpose of the requirements is to achieve safe operation of electrical power installations and a safe working environment in the vicinity of these installations. The measures (tests, measurements) stipulated in the standards are a contribution for avoiding failure in operations and cost-saving procedures. However – these measures can only be a snap-reading method for the installation. A properly planned concept of a preventive maintenance is only made possible with permanent monitoring.

### 6.5.2 Maintenance Terminology

In clause 7, the European standard DIN EN 13306 (DIN 31051):2001-09 on "maintenance terminology" deals with "Maintenance types and strategies". Maintenance is here defined as " Combination of all technical, administrative and managerial actions during the life cycle of an item intended to retain it in, or restore it to, a state in which it can perform the required function." Maintenance management includes all activities "that determine the maintenance objectives, strategies, and responsibilities and implement them by means such as maintenance planning, maintenance control and supervision, improvement of methods in the organization including economical aspects". The standard identifies various maintenance types and strategies. **Figure 6.4** gives a maintenance overview.

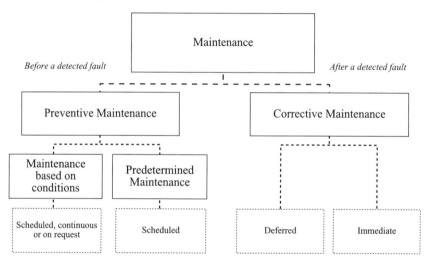

**Figure 6.4** Maintenance Overview according to DIN EN 13306, Annex A

### 6.5.3 Maintenance Strategy in IT Systems

The strategy of predictive maintenance according to the European standard DIN EN 13306 is conditioned based and "carried out following a forecast derived from the analysis and evaluation of the significant parameters of the degradation of the item (7.5)". Furthermore, the strategy of remote maintenance allows the "maintenance of an item carried out without physical access of the personnel to the item (7.7)".

By way of insulation monitoring and selective fault-location, the IT system may be applied as a remote maintenance strategy. The total insulation of the installation can be measured by insulation monitoring devices, current insulation measures are indicated, which allows trends to be calculated. Those measurements together with the fault location can be transmitted online via gateways to the appropriate places. There the actuating times and values of the insulation monitoring devices can be reset or other measures stipulated and hence increase maintenance effectiveness.

## Literature

[6.1] Wessel, W.: Feuergefährdete Betriebsstätten. Der Elektromeister. de-Sonderheft: Brandverhütung in elektrischen Anlagen. Heidelberg: Hüthig+Pflaum-Verlag, 1977, S. 28

[6.2] Brand, W.; Faller, E.: Elektrische Energieversorgung im Felde. Wehrtechnik (1993) H. 7, S. 52

[6.3] DIN EN 13306:2001-09: Maintenance Terminology; Trilingual version EN 13306:2001, Beuth Verlag GmbH, 10722 Berlin, Page 35

# 7 Applications of IT Systems

## 7.1 IT Systems in the Mining Industry

Early electrical engineering had already recognized that unearthed systems are very suitable for the special safety requirements in the mining industry [7.1], because a) the first ohmic insulation fault does not cause a break-down of operation and b) the fire-risk may also be reduced [7.1].

These systems are relatively small, hence the risk of shock currents is very low. In order to secure these advantages for the operation of electrical installations, it is important to adhere to the safety regulations, operating directions and occupational safety and health regulations.

Since the elaboration of the first standards in 1894 it became obvious, that not all aspects could be incorporated into one standard, and that standards with time had to be revised, modified and extended. For this reason, right from the beginning, special requirements for special locations, areas and industrial branches were created according to the state of the art of technology.

The first standards for underground mining with areas subject to explosion hazards were published as early as 1903 as "Safety regulations for the construction of electrical power installations". They were based on the principles for the rating of fire safety of electrical installations by the "Association of Private German Fire Insurance Companies". According to their contents and objectives, the electrotechnical standards were classified into three stages:

1) The first standards at the beginning of the century contained statements for both the erection and construction as well as the characteristics of materials to be used.

2) About 30 years later the insight was gained, that it was impossible to include all aspects in one standard. In 1938 a standard was published in Germany – DIN VDE 0118 – that only dealt with regulations on erection and was addressed just to fitters and installers of electrical installations in the mining industry, with modifications in 1941 and 1944 and general revisions in 1960 and 1969.

3) An attempt to regulate the fire and explosion hazards to people was also made in 1903 with the regulations for the erection of electrical power installations, with emphasis on the following:
   - earthing of exposed-conductive parts, "so that this part cannot carry a hazardous voltage which endangers persons who are standing uninsulated"

- measures so that electrical equipment cannot cause fire hazards
- permissible current loading for wires
- determination of the insulation status
- demands for explosion-proof machines and explosion-proof enclosed switches

## 7.1.1 Protective Measures in Underground Mining

The following chapter deals with the protective measures in underground mining according to the German national standard DIN VDE 0118, which has no equivalent to an international standard. However the author considers that the content of this standard may be interesting for the reader.

The special consideration required for the protection of persons and installations as well as underground mining against fire and explosion hazards necessitated the regulations for protective measures, exceeding normal standards for protective measures. Fires in underground mines are particularly dreaded, because the combustion gases of even small local fires may travel through different mine shafts and pits endangering a great number of people.

Only flameproof electrical machinery, equipment and fixtures for electrical locations and their direct working environment are specified as measures for fire-prevention. Electrical equipment with flammable insulating liquids may only be conditionally used. Outer coatings of wires have to be flame-retardant, the conductors current-carrying capacity has to have a set-value in order to avoid excessive heating. Other measures, such as the application of earth-fault neutralizing devices are only pointed out.

Sparks and electric arcs may become the ignition-source for high-explosive methane/air mixtures – also called pit gases. In order to combat such hazards the German standard DIN VDE 0118 requires that electrical installations are to be erected in such a way that they can be disconnected as soon as the critical threshold of the pit gas is reached.

The early requirements of the standard according to the specific regulating authorities did not sufficiently cover all aspects and hence more recommendations and stipulations, including protective measures, were issued in 1957. But the advancement of technology in the seventies made further revisions of the respective standards and regulations necessary. The standardization committees came to the conclusion that the various regulations should be combined again under one standard, but with different parts. The new comprehensive DIN VDE 0118 parts 1 – 3 was published in September 1990 with the title: "Erection of electrical installations for underground mining":

Part 1: General requirements

Part 2: Additional requirements for power systems

Part 3: Additional requirements for telecommunication systems

The latest publication of the series of standards was in 2001, with the general revision of all parts in cooperation with all relevant authorities, companies and operators. The following clauses are a selection from Part 1, with relevance to insulation monitoring.

### 7.1.2 Standards for Underground Mining

#### 7.1.2.1 DIN VDE 0118-1 (VDE 0118-1):2001-11

(Translation of the original German standard):

**Erecting of electrical installations in underground mining**

– Part 1: General requirements

*3. Definitions*

*3.20 Insulation monitoring device (IMD):* Electrical devices for monitoring the ohmic resistance of a system against earth.

*3.34.6 Protective conductor systems in underground mining*

*Systems with nominal voltages up to 1000 V and systems with nominal voltages exceeding 1 kV with protective measures against direct contact with the following characteristics:*

- *The system has no direct connection between live conductor and earthed parts; exposed-conductive parts are earthed via a protective conductor*

  *NOTE: For systems with nominal voltages up to 1000 V this requirement meets the requirements of an IT system according to IEC 60364-1*

- *The protective conductor is part of the entire system on all voltage levels and is only connected to an earthing system above ground*

- *Arrangement and cross-section of the protective conductor in wires are subject to special requirements, which, beside the consideration of the protection against the occurrences of high touch voltages, consider the requirements of fire and explosion protection*

- *Each galvanically separated system usually consists of an insulation monitoring device, which may cause interruption of the supply or an alarming action.*

*13. Protective conductor systems in underground mining*

*13.1.4 Deviating from 13.1.1, the following devices may be connected between the neutral point and the protective conductor:*

*b) Measuring devices for insulation monitoring or disconnection of earth faults with an a.c. internal resistance of at least 250 $\Omega/V$, yet at least 15 k$\Omega$ in systems with a nominal voltage up to 1000 V.*

## 13.2 Insulation monitoring

**13.2.2** Insulation monitoring devices shall be in compliance with the requirements 13.2.2.1 to 13.2.2.6. For remote control and signalling systems, clause 13.4.4 of DIN VDE 0118 applies to pit and slope conveyor systems.

**13.2.2.2** In systems with nominal voltages up to 1000 V, the drop in the insulation resistance of the system to be monitored below 50 $\Omega$/V system voltage against earth, shall be permanently indicated by a flashing light at the mounting location. When installing the insulation monitoring device within electrical operating areas or within closed operating areas, the flashing light may be omitted if an automatic indication to a permanently manned station is ensured.

**13.2.2.3** The setting of the response values of insulation monitoring devices shall only be made possible after opening the enclosure.

**13.2.2.4** The response values shall be directly identifyable, without any conversion factor, at a permanently fixed scale or similar facility.

**13.2.2.5** The insulation monitoring device shall be equipped with a test device capable of testing the functions with external-testing functions

**13.2.2.6** In systems with nominal voltages up to 1000 V, insulation monitoring devices in accordance with IEC 61557-8 shall be applied.

## 18 Protection of wires against the dangers from mechanical influences

**18.2.3** To comply with requirements from 18.2.2.4 and 18.2.2.5 the application of insulation monitoring devices according to IEC 61557-8, residual current protective devices according to IEC 61008-1 and IEC 61008-2-1 or residual current relays in connection with switching devices is required.

**18.3.2.2** Circuits shall be monitored with an insulation monitoring device, which automatically disconnects the system on the condition of low ohmic insulation fault to earth or low ohmic insulation fault to exposed-conductive parts after at least 1.5 s.

In systems with functional extra-low voltage according to 15.2 [Note of the translator – may only be used in systems with nominal voltages up to 50 V a. c. or 120 V d. c.]

**18.4.4.4** Auxiliary circuits shall consist of an insulation monitoring device, capable of indicating at least a low ohmic insulation fault.

## 19 Electrical protective installations and associated circuits

**19.1.2** Electrical protective installations according to nominal voltage shall include at least the following equipment:

**19.1.2.1** In systems with nominal voltages up to 220 V equipment for disconnection at:
- Short-circuit between live conductor and protective conductor
- Interruption of the monitored circuit for the protective conductor

- Low-ohmic insulation fault, provided the earth fault is switched off after 1.5 s by the insulation monitoring device according to 13.2

**19.1.2.2** In systems with nominal voltages above 220 V and up to 1000 V without converter equipment for disconnection at:
- Short-circuit monitoring conductor/protective conductor
- Interruption of the monitored circuit for the protective conductor
- Low ohmic insulation fault. If the insulation resistance is reduced to 20 k$\Omega$, at least the part of the circuit which is being monitored by the protective equipment shall be disconnected after at least 1.5 s; the disconnection may be made by the insulation monitoring device according to 13.2. If the low ohmic insulation fault is complete, this part of the system shall be switched off after at least 200 ms
- Low ohmic insulation fault, provided the insulation fault will not be disconnected by the insulation monitoring device according to 13.2.
- Short-circuit between outer conductor and monitoring conductor, if the monitoring conductor is concentric according to 19.2.2.2

**19.1.2.3** In systems with nominal voltages above 220 V and up to 1000 V with converter installations for disconnection according to 19.1.2.2. However if, in the case of an ohmic insulation fault, the insulation resistance decreases below 20 k$\Omega$, the circuit being monitored has to be de-energized by the protective device, within the quickest time possible for the measured-value acquisition. Conductor monitored sub-circuits, which are not fed by the converter, must be de-energized after at least 200 ms in the event of a low ohmic insulation fault.

## 7.1.3 Protection Against Electric Shock in Underground Mining

The protective measures in underground mining explained in the previous chapter are stipulated requirements. Such systems have no direct connection to earthed parts. All exposed-conductive parts are earthed via a protective conductor, which is connected to an earthing system above ground. Layout and cross section of the conductor are subject to defined requirements. Each galavanically separated system is generally equipped with a protective device, for example an insulation monitoring device, which provides protection by disconnection or indication. In practice, an additional equipotential bonding in underground mining is everywhere: all exposed conductive parts of electrical equipment and – because of the constructive assembly and the combined effects – also other exposed-conductive parts are connected to the protective conductor in a durable manner and sufficiently conductive through all voltage levels.

The protective concept in underground mining can be described as follows:
- In systems with nominal voltages up to 1000 V a.c., the condition of the insulation shall be monitored and a signal shall be released, when the insulation resi-

stance falls below a predetermined value. For this purpose insulation monitoring devices in accordance with IEC 61557-8 are generally applied.

- Conductors in mobile electrical equipment used in mining, for example in a drum-type cutter-chain-loader, shall be monitored by an insulation monitoring device, which automatically disconnects the system in the event of an insulation fault to earth or to exposed-conductive parts after 1.5 s, if they are operated without an electrical device in accordance with DIN 0118, clause 9 (see the above chapter).
- Systems with nominal voltages above 220 V up to 1000 V a.c. shall be designed in such a way that in the event of a single low ohmic insulation fault to earth, at least the faulty part is disconnected after at least 1.5 s. In systems with converters the faulty system part shall be disconnected within the shortest practicable delay.
- Reclosure shall be prevented via an earth-fault lock-out device, as long as the unacceptable insulation fault exists.

Earth-fault lock-out devices are applied to monitor insulation faults, for example in disconnected motor supply conductors, to prevent reclosure, as long as the predetermined insulation value is below a minimum value.

Lock-out devices of this kind are often integrated in insulation monitoring devices. The voltages of their measuring circuits shall comply with special requirements.

## 7.2 IT Systems with Insulation Monitoring on Board Ships

The variety and the number of electrical and electronic equipment on board of sea-faring ships and off-shore facilities (marine/oil) is steadily increasing and with it the requirements on electrical installations to protect the crew from electrical dangers and to guarantee reliable and safe electrical supply. Principle considerations start with the selection of the appropriate system type. Many international standards recommend the unearthed IT system. The following explains the special characteristics of this system type with appropriate insulation monitoring.

### 7.2.1 Standards and Regulations

There are many standards and regulations that govern the construction and operation of sea-faring ships, which significantly influence the design of the electrical equipment:

- International Shipping Safety Treaty
- Agreements for the safety of (human) life at sea (Solas)
- IEC 60092-101, Ed. 4.1:2002-08, Electrical installations in ships – Part 101: Definitions and general requirements[1]
- Requirements of classification bureaus
  - Loyd's Register of Shipping
  - Det Norske Veritas
  - American Bureau of Shipping (ABS)
  - Canadian Coast Guard
  - International Maritime Organization
  - Rules for the Classification of Construction of Sea-going Ships (GUS)
  - Bureau Veritas, Paris
  - American Society for Testing and Materials (ASTM)

Common objective of these treaties, agreements and requirements is the achievement of a high degree of safety and reliability by standardizing characteristics, data and measurements. The planning and design of electrical installations in the operational field of "sea-vessels" require the consideration of several aspects:

- A ship is a self-sufficient system, able to remain independent of external supply over a prolonged period of time
- Climatic demands range from tropical to arctic
- Specific demands on high sea through list, shocks (for example through ice)
- Chemically aggressive influence of sea-water
- Change of location/positions (different supply systems ashore)

---

[1] The whole series of this standard with the general title "Electrical installations in ships" is listed in chapter 17

### 7.2.2 Permissible System Types Onboard Ships (Marine/Oil)

Electrical supply systems onboard ships and off-shore facilities are classified into primary and secondary systems. While primary systems are directly connected to the generator, the secondary systems have no direct connection. For example they are insulated from the generator by a transformer with two windings or by a motor-generator or converter set.

The principal lay-out of such systems and installations shall be such that:
- Supply, essential for safety, will be maintained under various emergency conditions
- The safety of passengers, crew and the vessel from electrical hazards is ensured
- The requirements of the various regulations and standards are adhered to

### 7.2.3 TN and IT Systems on Board Ships

Unlike earthed systems, called TN or TT systems, who have a direct connection to earth, the unearthed IT systems have no connection of live conductors to earth. Some system types applied on ships like three-phase IT and TN systems, as well as single-phase and d.c. systems, with respective insulation monitoring device are shown in **Figure 7.1**.

The IT system is supplied by a transformer, a generator, accumulator or another independent power source. Since there is no direct connection between any live conductor and earth, a first insulation fault will not influence the function of the connected electrical equipment. Hence, no disruption of operation will occur.

Continuously monitoring the insulation resistance guarantees high reliability of the IT system. It is a requirement, which is also included in the regulations and standards. **Figure 7.2** shows an insulation monitoring device licensed by the Germanische Lloyd.

The following is a quotation from IEC 60092-201:

"Insulated distribution systems, when a distribution system, whether primary or secondary, for power, lighting or heating, with no connection to earth is used, a device capable of continuously monitoring the insulation level to earth and of giving an acoustic or optical indication of low insulation values shall be provided. See IEC 60092-502."

"For insulated distribution systems with nominal voltage exceeding 500 V, consideration should be given to installing a device or devices which continuously monitor the insulation level and give audible and visual alarm in case of abnormal conditions."

Excerpt from a standard of the "Institute of Electrical Engineers" (Great Britain) "Regulations for the Electrical and Electronic Equipment of Ships, ed. 1972":

"Where an earth-indicating system using either two or three lamps, as appropriate,

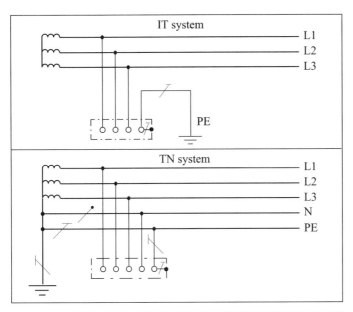
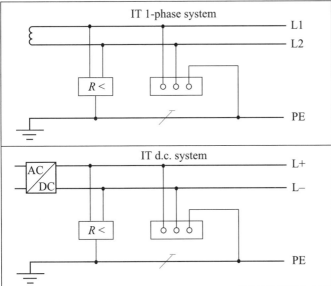

**Figure 7.1** Diagram of system types on ships

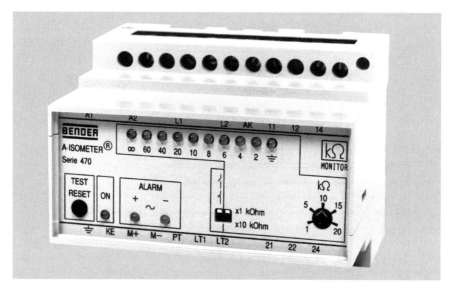

**Figure 7.2** Recommended insulation monitoring device type IR470LYX approved by Germanischer Lloyd - GL (Photo supplied by Bender, Germany)

is adopted, earthing lamps should be of metal-filament type each not exceeding 30 watts. The system employing a single lamp should not be used. To facilitate comparison of the brilliance of earth-lamps, they should be of clear glass and should be placed not more than 150 mm apart."

This aforementioned kind of insulation monitoring dates back to 1885. However today, in view of the availability of state of the art technology, like insulation monitoring devices, the standard is no longer practical. The adaptation of a more currently relevant standard is necessary. Rather amazingly is the fact, that the outdated three-voltmeter-method with filament lamps is still recommended in the standards.

ASTM (American Society for Testing and Materials) has decisively advocated a modernization. Three standards have already been published. The first, F 1207M-96 (Re-approved 2002) is titled "Electrical Insulation Monitoring for Monitoring Ground Resistance in Active Electrical Systems". The second standard F 1134-94 (Re-approved 2002) is called: "Insulation Resistance Monitor for Shipboard Electrical Motors and Generators". The title of the third standard F 1669M-96 (Re-approved 2002) is called: "Standard Specification for Insulation Monitors for Shipboard Electrical Systems". (See also Chapter 10.3)

**Summary**

IT systems are frequently applied on ships and off-shore facilities, especially in areas where a high degree of safety is vital. A transformer or a motor-generator are normally part of a secondary systems; hence the prerequisite for an IT system are available, without the need of further isolating transformers [7.2 to 7.12].

The basis for a safe and reliable distribution system is the application of IT systems with supplementary equipotential bonding, insulation monitoring devices, complemented with insulation-fault-location-equipment. On condition that the standards and requirements are applied, damages to persons or equipment – caused by electrical accidents or fires – can be kept at a minimum.

### 7.2.4   IT Systems in Ships of the German Navy (Bundeswehr/Germany)

By kind permission of the "Bundesamt für Wehrtechnik und Beschaffung" (Defense Technology and Procurement Office) in Koblenz, some excerpts of the "Building Regulation (BV) for ships of the German Federal Armed Forces" are quoted below, see also **Figure 7.3** and **Figure 7.4**).

A comprehensive overview is given in the entire building regulation BV 30. The

**Figure 7.3** Frigate - series F122 of the German Federal Naval Forces

**Figure 7.4** Control centre of the frigate F122 with insulation monitoring panel

BV 30 applies to ships of the Federal Naval Forces. This regulation is an aid for the planning of electrical installations with general guidelines and bears the title: "Building regulations for ships of the German Federal Naval Forces 3000-1 – Electrical installations and planning and general guidelines for surface vessels, edition 03.90." (Bauvorschriften für Schiffe der Bundeswehr 3000-1 – Elektrische Anlagen und Planung und allgemeine Richtlinien für Überwasserschiffe, Stand 03.90).

The building regulation includes design, construction and production guidelines.

The following requirements exclusively refer to the type of distribution system and insulation monitoring.

The general requirements of voltage systems are:
- All main and sub-systems have to be designed for all-pole disconnection and must be all-pole isolated (neutral point unearthed). All shipboard systems shall be designed as radial systems with selective short-circuit protection. As distribution system, the IT system in accordance with IEC 60364-4-41 is applied.
- Protection against indirect contact

  For all ships of the German Federal Naval Forces the protective measure against indirect contact "Protection through disconnection and indication" according to

IEC 60364-4-41 shall be applied, where in particular the permissible protective measures appropriate to the type of distribution system required for ships of the German Federal Naval Forces apply. Insulation monitoring devices shall be installed.

- Insulation monitoring

  In main systems and sub systems, which are galvanically separated from these systems, stationary insulation monitoring systems shall be installed. They shall be capable of continuously monitoring the insulation state of all conductors against hull or the protective conductor.

  The monitoring devices shall not be provided with switch-off functions unless this is required for a special application explicitly stated in the specification.

The devices shall be designed in a way that the minimum values indicated in IEC 61557-8 are adhered to. The response values of the insulation monitoring devices shall be adjustable so that they can be adapted to the permissible insulation resistances indicated in the specifications valid for the respective system.

If the permissible insulation resistance falls below the minimum value, visual and acoustical warning signals shall be send to the electrical control board in the master control station of the vessel.

## 7.3 IT Systems with Insulation Monitoring in Railway Applications

Future rail traffic calls for innovative solutions. The steadily increasing requirements for effectiveness, economic efficiency, reliability and above all operating safety require the use of highly sophisticated traction technology and power system protection. This is an ideal application field for unearthed IT systems with insulation monitoring. Numerous international standards are available for railway application. Just to mention a few:

- IEC 60077 Parts 1 to 5: Railway application, Electric equipment for rolling stock
- IEC 60310:2004-04, Traction transformers and inductors on board rolling stock
- IEC 60850:2000-08, Supply voltages of traction systems
- IEC 61991:2000-01, Rolling stock – Protective provisions against electrical hazards

### 7.3.1 Examples of Applications for IT System with Insulation Monitoring

There are two applications in rail vehicles, where IT systems are preferred to earthed systems. **Figure 7.5** shows the structure of the power supply system of a locomotive engine with two typical IT systems and insulation monitoring functions.

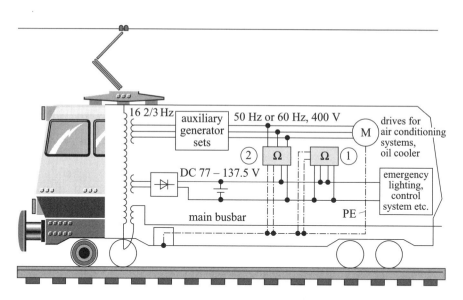

**Figure 7.5** Design of a power supply system in a locomotive engine

**Figure 7.6** Heathrow Express

## 7.3.2 Application Fields for IT Systems with Insulation Monitoring

Meanwhile many IT systems with insulation monitoring devices are finding application fields nationally and internationally. In many railway projects accumulator-backed d.c. systems are being monitored successfully with insulation monitoring devices. Other applications, like the monitoring of external charging stations or drives in the engine, can be easily realized with specially modified versions of the devices. Auxiliary converters, like the ones in the Heathrow Express for example (**Figure 7.6**), which supply air-conditioning systems, lighting systems and so on are now often applied as insulated systems. The increasing demands on the rail traffic require sophisticated technology. That is the reason why today more and more systems are designed as IT systems with insulation monitoring. The number one argument for it is the high reliability of this system type.

## 7.3.3 Requirements on Insulation Monitoring Systems

Insulation monitoring devices used in IT systems for rolling stock shall comply with the requirements of DIN EN 50155 (VDE 0115-200):2004-09. The measuring method used in an insulation monitoring device must be adapted to the respective application. Due to steadily changing system constellations and applications of converters, only very few measuring principles like the AMP-measuring principle for example, are able to reliably determine the insulation resistance. Choosing the wrong measuring method may lead to measurement errors and hence to false alarms. The applied measuring principle may be unable to determine the insulation resistance because of interference superimposed onto the system. Thus, a developing earth fault may remain unrecognised.

Required insulation resistances for accumulator installations are stipulated in DIN VDE 57510 (VDE 0510):1977-01. Requirements regarding the insulation resistance in a.c. or three-phase systems are stipulated in DIN EN 50122-1 (VDE 0115-3):1997-12 and in DIN VDE 0105-100 (VDE 0105-100):2005-06. To facilitate the setting of the insulation monitoring device, **Table 7.3** lists recommended and required insulation resistance values.

It may be sensible to supply insulation monitoring devices by battery-backed d.c. systems, in order to be able to monitor the de-energized system. Since the location of application naturally frequently changes, the insulation monitoring device should be equipped with external test and reset buttons. In individual cases the connection of an external measuring instrument for the indication of the insulation resistance is provided. Using the IT system with insulation monitoring makes it easier to realize some clauses of DIN EN 50155 (VDE 0115-200):2004-01, Bahnanwendungen – Elektronische Einrichtungen auf Schienenfahrzeugen (Railway applications – Electronic installations in rolling stock). Because of continuous monitoring of the insulation resistance, for example, exceeding the value of the insulation resistance is indicated and the fault can be cleared without delay. This eases the

| Standard | Nominal Voltage | Required Insulation Value | | | | Recommended Response Value of the Insulation Monitoring Device | |
|---|---|---|---|---|---|---|---|
| | | | | | 1) | | 1) |
| DIN VDE 0105-100 (VDE 0105-100):2006-06 Electrical installations in normal service | AC 230 V | 50 Ω/V | 11 kΩ | 300 Ω/V | 69 kΩ | 17 kΩ | 104 kΩ |
| | AC 277 V | 50 Ω/V | 14 kΩ | 300 Ω/V | 83 kΩ | 21 kΩ | 125 kΩ |
| | AC 400 V | 50 Ω/V | 20 kΩ | 300 Ω/V | 120 kΩ | 30 kΩ | 180 kΩ |
| | AC 480 V | 50 Ω/V | 24 kΩ | 300 Ω/V | 144 kΩ | 36 kΩ | 216 kΩ |
| | AC 500 V | 50 Ω/V | 25 kΩ | 300 Ω/V | 150 kΩ | 38 kΩ | 225 kΩ |
| | AC 690 V | 50 Ω/V | 34 kΩ | 300 Ω/V | 207 kΩ | 51 kΩ | 311 kΩ |
| DIN EN 57510 (VDE 0510): 1977-01 for Accumulators and battery systems | DC 24 V | 100 Ω/V | 3 kΩ | | | 4 kΩ | |
| | DC 48 V | 100 Ω/V | 5 kΩ | | | 8 kΩ | |
| | DC 72 V | 100 Ω/V | 8 kΩ | | | 11 kΩ | |
| | DC 96 V | 100 Ω/V | 10 kΩ | | | 12 kΩ | |
| | DC 110 V | 100 Ω/V | 12 kΩ | | | 17 kΩ | |

**Table 7.3** Required insulation values and recommended setting of the response value of insulation monitoring devices
1) Insulation values for well-maintained installations

demand on the electronic equipment and more reliable assessments may be achieved. In addition, for maintenance reasons, the system is designed without the need for periodic maintenance. Furthermore, the operator in accordance with DIN EN 50155 (VDE 0115-200) may ask for diagnostic-equipment, in order to assess the state of the supply system. Some clauses of this standard may be easier achieved by the application of insulation monitoring devices in the IT system. Requirements from clause 4.6 (automatic test devices), can be fulfilled with selective insulation-fault-location devices.

### 7.3.4 Accumulator-Backed Safety-Oriented D.C. System

Accumulator-backed d.c. systems are safety-oriented, which means that in the event of a fault, the emergency power supply is maintained for a time. The task of the system is to continue to supply the safety-relevant equipment, for example emergency lighting, door and brake control systems. After that, depending on the need, individual consumers are disconnected in a certain time grading. Since this system is able to supply a number of wagons, depending on the length and construction of the train, it is not quite possible for that reason to define the electrical transients. An intelligent measuring process is necessary to monitor the insulation. Insulation monitoring devices with integrated AMP measuring process are therefore often applied in this application field. See chapter 11.4.1 for a closer description of this process. The insulation monitoring devices are being supplied by the monitored system. In this way monitoring the insulation of accumulator-backed d.c. systems is ensured, even in the event of a power failure. Since in this system safety and availability of the supply are a top priority, the IT system is the best alternative for insulation monitoring, for reasons already described in the previous chapters.

### 7.3.5 Converters in Main Circuits

Different variations of converters for different functions are being applied in train engines. Auxiliary converters for example supply the air-conditioning, lighting, accumulator charging systems and the control circuits. One task of the converter is to convert the frequency of 16 2/3 Hz into 50 Hz in order to use standard devices with compotitive prices. Controlling the traction drives with different power inverters is another function, which is described in IEC 60349-1, "Electric traction – Rotating electrical machines for rail and road vehicles – Part 1: Machines other than electronic convertor-fed alternating current motors". Definitions of the different converter types may be found in DIN EN 50207 (VDE 0115-410):2001-03, "Electronic power converters in railway vehicles" and IEC 61287-1:2005-09, "Railway application – Power converters installed on board rolling stock – Part 1: Characteristics and test methods". The definition for an electric power converter is:

"Device for changing one or more characteristics associated with electric energy.

Note – Characteristics associated with energy are for example voltage, number of phases and frequency including zero frequency" [IEV 811-19-01 mod.]

The international standard IEC 62236-5:2003-04, "Railway applications – Electromagnetic compatibility – Part 5: Emission and immunity of fixed power supply installations and apparatus" gives guidelines for electromagnetic compatibility. For EMC-requirements, mains filters (capacitors) are increasingly applied. To add parallel connections of several converters are used. Because of the parallel connection of the filters and system leakage capacitances, the insulation resistance is reduced considerably.

A further reduction of the insulation resistance can be recognized and indicated by the IT system by way of insulation monitoring. The operator is then able to clear the fault with the shortest practical delay, in order to avoid the build-up of fault currents (see IEC 60364-4-41:2001-08, 413.1.5.4 note 1). This allows sensitive systems to be protected against power failure and maloperation caused by insulation resistance faults or earth faults. Insulation monitoring devices, which measure reliably in converter circuits, usually work with the special intelligent AMP measuring method (see chapter 11.4.1). The recommendation here also applies to supply insulation monitoring devices from the accumulator-backed d. c. system (see **Figure 7.5**), and to monitor de-energized systems. This allows the detection and indication of insulation resistance faults or ohmic insulation faults. This gives the operator the chance to clear the fault swiftly. Any unnecessary delays through repairing damaged or destructed systems can effectively be avoided in this way.

## 7.4 IT Systems in Electric Vehicles

Electric vehicles are automobiles (limousine, truck, bus), which are not driven by combustion engines, but are completely or partly electrically driven.

There are different concepts regarding electric drives, of which some are still under development, others have proved themselves:

- the accumulator-supplied electric vehicle
- the diesel-electric system
- electric-vehicles with fuel cells
- hybrid driving gear, for example diesel electric drives plus accumulator plus electric engine, or gas turbines plus accumulator plus electric engine

Rather unnoticed by the public at large and a little inconspicuous, in recent years some companies have developed, tested and marketed electric vehicles around the globe. Rising environmental pollution and oil-dependency are the main causes for the re-awakened interest. Environmental legislation foster the interest in alternative technology as well as progressive environmental ministries, like the one in California, environmental consciousness and conservation will all contribute to an in-

**Figure 7.7** Electric vehicle fleet on Ruegen [7.1]

creasing production of vehicles without polluting emissions. This is the so called Zero Emission Vehicle (ZEV). In California approximately 10 % of vehicles will conform to those demands by 2003.

Germany has not had those demands yet. But research, development and testing in this field is well on the way. Field tests involving electric vehicles have been conducted as early as 1992 on the isle of Ruegen, at the German Baltic-Sea [7.15]. Over several years, 60 electric vehicles from different manufacturers were tested in every-day use. The load current was partly supplied by solar power and wind energy (**Figure 7.7**).

At the early 1990s La Rochelle in France also tested a fleet. The La Rochelle Council, the French national power supplier EdF and the multinational automobile group PSA Peugeot-Citroen made a field study with 50 electric vehicles. At the end of the testing period in 1995 the vehicles combined drove 700.000 km.

Europe wide 10.000 electric vehicles were registered in 1999. In 1995 approximately 1700 electric vehicles took part in tests all over Europe. [7.13]

There are many fleet tests all over Europe: Belgium, Spain, The Netherlands, Sweden, Finland, Norway, Switzerland, Italy, Austria, Great Britain, France and Germany are all part of it. Besides, the idea of electrically driven cars is nothing new – the early 20ies already saw electrically driven cabs, in the 1950ies there were trucks with electrical engines and Great Britain has about 20000 electrical milk delivery vans on the road every day for years. [7.17]

The results of the fleet tests were, that the interest in electrically operated road vehicles is as great as never before. Many Governments view it as one of the most important drives in the vehicle of the future. The successful application of serial-production and pre-serial-production vehicles proves that a great number of electric vehicles are regarded as being able to withstand the day-to-day demands of traffic. The high rate of user satisfaction of the La Rochelle trial should be noted! [7.14]

Testing the acceptance had been a part of many of the practical oriented tests. As a rule, the vehicles were accepted after a phase of familiarization. Only a low percentage of test persons said, that they no longer wished to drive an electric vehicle [7.17].

The aspect of environment protection is especially effective in the solar and electric concept (solar+E-concept). This combines the use of light and energy-efficient electric vehicles with their environmental-friendly power supply through regenerative energy sources. Regenerative energy is gained from the sun, the wind and water, as well as biomass. The German Federal Solar-Mobile Association (Bundesverband Solarmobil bsm) describes the concept as follows:

"Solar mobiles are electric light vehicles (ELV). They receive their energy supply from an interconnection between environmental-friendly and regenerative energies (sun, wind, water), the "fuelling" station is every 230 V household socket-outlet."

Extracting environment-friendly produced electric energy, which had been fed into the public power supply system, from a 230 V socket-outlet, is termed "solar interconnection".

Modern solar and electric mobiles have an energy consumption of approximately 10 kWh per 100 km and a mean requirement of energy of about 5 kW. Due to their mode of drive they are termed electric mobile. In order to preserve the $CO_2$ level, the vehicle owner has to see to it, that the necessary load current is produced environmentally friendly. [7.13]

It is quite interesting reading the testimonials of private institution and private individuals. There is much enthusiasm in the pursuit and publication of the idea of environment-friendly vehicles. [7.16]

### 7.4.1 Protective Measures in Electric Vehicles

In conventional vehicles with combustion engines the distribution system as a rule is 12 V d.c. or 24 V d.c. For this reason it is below the permissible threshold of the touch voltage, which has been set at 50 V a.c. and 120 V d.c. by international agreements.

This is not the case with electric vehicles. Depending on the different drive concepts, as mentioned before, distribution systems ranging from 100 V up to 600 V are quite common. This represents dangerous dimensions resembling the electrical systems of industry and households. But what is the difference between the distribution board of an electric vehicle and that of an industrial installation? The answer in terms of the required protection has to be: NOTHING!

Inevitably, designer, manufacturers and drivers are confronted with terminology like:

- Supply system configuration
- Protective measures
- Electrical safety
- Protection against electric shock
- Touch voltage
- Preventive maintenance
- Operational safety
- Insulation fault
- Leakage current
- Fault current

The European standard-series DIN EN 61851 (VDE 0122), titled "Electric vehicle conductive charging system" deals with vehicles with electric drives over several parts (**Table 7.4**):

| European | German | International | Title |
|---|---|---|---|
| DIN EN 61851-1:2001-11 | VDE 0122-1 | IEC 61851-1:2001-01 | Part 1: General requirements |
| DIN EN 61851-21:2002-10 | VDE 0122-2-1 | IEC 61851-1:2001-01 | Part 21: Electric vehicle requirements for conductive connection to an a.c./d.c. supply |
| DIN EN 61851-22:2002-10 | VDE 0122-2-2 | IEC 61851-1:2001-01 | Part 22: AC electric vehicle charging station |

**Table 7.4** Standard-series for electric vehicles

### 7.4.2 Distribution Systems of Electric Vehicles

Basically all types of supply systems are suitable as distribution systems in electric vehicles:
- IT system
- TN system
- TT system

In practice for reasons of high availability mainly IT systems are found. Disconnection of the distribution system while driving is completely unacceptable and must be avoided. It is only sensible to disconnect only the part of the distribution system, which has the insulation fault.

The IT system of an electric vehicle can be quite complex through the regulation of the drive and through interference suppressors, from the technical-measurement point of view. The selection of the appropriate insulation monitoring devices is therefore of great importance. The AMP-measuring principle, described in chapter 11.4.1, (Microprocessor-controlled AMP measurement process for the general application in a. c. and d. c. IT systems), has proven to be very suitable for electric vehicles.

Selective disconnection of the part of the distribution system on which the insulation fault is occurring has shown to be feasible. The insulation-fault-location system required for this is readily available on the market. For more detailed information on this subject see chapter 10.5, Insulation-fault-location system in a. c. and d. c. IT systems.

National and international standard institutions are aware of the problem. The European standard DIN EN 61851 (VDE 0122-1) (see **Table 7.4**) is about the electrical equipment of electric road vehicles with special emphasis on the charging method. There are normative and informative references to international standards:
- IEC 60038, consolidated edition: 2002-07, IEC Standard Voltages
- IEC 60364-4-41, ed.5.0:2005-12, Low-voltage electrical installations – Part 4-41: Protection for safety - Protection against electric shock

Like in all electrical systems, coordination between system type and protective device, is also essential for electric vehicles. To be taken into consideration are the design of electrical distribution systems as well as the method of charging the accumulators. By analogy with the electrical distribution system of the industry, various solutions for realizing the protection and monitoring concept are also to be realized for the electric vehicles.

The analogy with the industry and private households shows that the wheel does not have to be reinvented over and over again. The protection and monitoring concepts of the electro-technical industry also offer electric vehicles sufficient protection. A reason to make good use of this advantage.

### 7.4.3 Recharging Stations of Electric Vehicles

Principally a distinction has to be made between "inductive" and "conductive" loads of electric vehicles.

With **inductive loads**, there is no conductive connection between charging system and the distribution system of the electric vehicle. In order to refuel, a pivot or paddle is inserted into a slot. The charging station converts the system voltage (e. g. 110 V a. c., 60 Hz) into a high-frequency alternating current, which reaches the vehicle via the paddle (**Figure 7.8**).

Both systems can be protected and monitored with respective measures through the electrical isolation between charging system and distribution system. The distribution system will receive its own insulation monitoring device. The charging station, as long as it is supplied through an earthed system, will receive both a residual current protective device (RCD) as protective measure and a residual current monitor (RCM) for preventive maintenance. Both systems are compatible. There will be

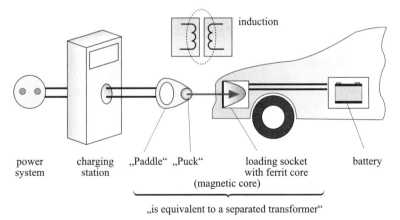

**Figure 7.8** Example of an inductive charging station

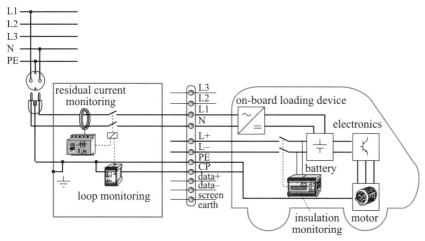

**Figure 7.9** Example of an conductive charging station

no interference neither between both systems, nor between the protective measures. Therefore, when considering and evaluating protective and monitoring measures it usually does not at all matter, whether the system type for the charging station is earthed or unearthed.

However, this is different with conductive loads of electric vehicles. The load current flows directly into the vehicle via a plug-in connector cable, hence a coupling occurs between charging system and the distribution system.

Since the conductive charging station can be both an earthed or unearthed system type, the user now has to make a decision regarding the coupling of the systems and the appropriate protective measures (**Figure 7.9**).

If the IT system of the distribution system is being coupled with the IT system of the charging station, two insulation monitoring devices are suddenly part of the system. This can lead to uncertainties in the measured values and hence to unsafe conditions, which means, that one of the insulation monitoring devices has to be disconnected.

If the IT system of a distribution system is being coupled with the TN (or TT) system of a charging station, the insulation monitoring device, because of the grounding of the charging station, will recognize the fault and will release. It is possible to activate the measured-value suppressor of the insulation monitoring device during the charging process, in order to prevent fault indication.

At the same time it must be guaranteed, that the protective and monitoring devices of the charging station will correspond with the possible direct currents from the distribution system. This is a problem, which has to be considered when planning and designing a charging station.

### 7.4.4 International Standard for Charging Stations of Electric Vehicles

The growing interest in electric vehicles is also recognized Internationally by the standard committees. The IEC has published the following standard series:

IEC 61851-1:2001-01

*Electric vehicles conductive charging systems.*

Part 1: General requirements

Part 21: Electric vehicles requirements for conductive connection for an a. c./d. c. supply

Part 22: a. c. electric vehicle charging stations

In part 1 insulation monitoring is required as an additional protection measure, when an unearthed charging system is used.

Standard activities on protective measures of charging stations for electric vehicles in The United States of America have resulted in the publication of UL 2231 by Underwriters Laboratories in 2002:

UL 2231-1: Personnel Protection Systems for Electric Vehicle (EV) Supply Circuits

UL 2231-2: Particular Requirements for Protective Devices for use in Charging Systems

These two standards contain particular details for the application of insulation monitoring devices.

#### 7.4.4.1 Insulation monitoring devices according to UL 2231

The requirements on insulation monitoring devices in the UL 2231 standards are clearly described. The following is an original quote:

##### 7.4.4.1.1 UL 2231-1, General Requirements

The general requirement of UL 2231-1 regarding insulation monitoring devices (the American term is isolation monitor) are as follows [7.18]

*2.16    Isolation Monitor/Interrupter – A device that monitors the insulation resistance of an isolated circuit to ground and prevents energization of the charging circuit or disconnects an energized charging circuit when insulation resistance drops below a predetermined value (see 5.3).*

*2.17    Isolation Monitor/Interrupter with self-check – A device similar to that described in 2.16 except that it is also equipped with an automatic supervisory circuit that periodically checks the operation of the isolation monitor and does not permit energizing the circuitry or, during operation, disconnects the energizing circuitry connected to the load terminals of the isolated circuit under conditions where the isolation monitor does not function properly.*

5.3     *Isolation monitor/interrupter*

5.3.1   An isolation monitor/interrupter shall monitor the insulation that provides the electrical isolation from ground of an isolated circuit. If the symmetrical as well as asymmetrical resistance to ground is reduced below a predetermined value, then the device shall not energize the circuitry connected to its load terminals. If it is of a type that monitors continuously, and the reduction in resistance occurs while the system is in operation, the device shall act to cause the supply to be disconnected during a charging cycle.

5.3.2   An isolation monitor can be on-board or off-board the vehicle or there can be both on-board and off-board monitors.

5.3.3   When only an on-board monitor is provided, it shall function to monitor the entire isolated circuit and shall have the capability of communicating to cause the external supply to be disconnected, in addition to disconnecting the batteries, if the resistance to ground is reduced below the set point.

5.3.4   When only an off-board monitor is provided, it shall function to monitor the entire circuit and shall have the capability of causing both the supply and the batteries to be disconnected if the resistance to ground is reduced below the set point.

5.3.5   When both an on-board and an off-board monitor are provided, and they are compatible in that they do not interfere with each other, they shall monitor the circuit and shall have the capability of causing the external supply and the batteries to be disconnected if the resistance to ground is reduced below the set point.

5.3.6   When both an on-board and an off-board monitor are provided, and they are not compatible in that the measuring circuit of one monitor would be interpreted as a loss of isolation by the other monitor, or when interaction between the monitors prevents both monitors from detection of lost isolation, then one of the monitors shall be disconnected and the remaining monitor shall have the capability of monitoring the entire circuit and communicating to cause the disconnection of the external supply and the batteries if the resistance to ground is reduced below the set point.

5.3.7   An isolation monitor/interrupter that is intended to operate on an isolated part of a circuit, for example a vehicle, and is designed to become grounded when it is plugged into a charger shall be designed to check the isolation of the isolated part of the circuit before charging begins, and shall not continue to function during the remainder of the charging cycle. For a circuit designed to operate as an isolated circuit, isolation monitoring shall continue throughout the entire charging cycle.

### 7.4.4.1.2 UL 2231-2 Supply Circuits

The requirements for Isolation Monitor/Interrupters are specified in UL 223-2 "Supply Circuits: Particular Requirements for Use in Charging Systems". The following is an original quote [7.19]:

*14      Isolation Monitor/Interrupter*

*14.1    An isolation monitor/interrupter shall monitor the resistance of the insulation that provides electrical isolation from ground of an isolated circuit. When the resistance from any ungrounded conductor to ground is less than 100 ohms/volt, based on nominal system voltage, the device shall not allow the load circuit to be energized.*

*14.2    The A.C. internal resistance of the monitor shall be at least 250 ohms/volt of nominal system voltage. The DC internal resistance of the measuring device shall be at least 30 ohms/volt of nominal system voltage and shall limit the measuring output current to 5 mA.*

*14.3    An isolation monitor that has an adjustable trip setting shall also have a means to prevent tampering with the setting.*

*14.4    An isolation monitor shall have a built-in test circuit or means to connect circuitry to verify correct operation of the monitor.*

*14.5    When the circuit is intended to operate as an isolated circuit throughout the entire charging cycle, then the isolation monitor/interrupter shall continue to monitor the circuit during charging and shall open the circuit when the resistance to ground from any conductor is less than 100 ohms/volt.*

*14.6    An isolation monitor intended to be used on systems where the voltage between conductors is greater than 150 Vrms and where only basic insulation is provided shall have an automatic self-test feature that:*
*a) tests the monitor to determine whether or not it is functioning within indended limits and*
*b) does not permit energizing the circuitry connected to the load terminals under the condition that the isolation monitor is not functioning properly.*

*14.7    The test mentioned in 14.6 shall be performed prior to each charging cycle or at regular intervals. If testing is performed at regular intervals, the test should be performed at least once each hour.*

*14.8    To demonstrate that an isolation monitor/interrupter meets the requirements of 14.1 – 14.5 test as described in System Test Requirements, Section 21, shall be performed.*

The attentive reader would have noticed that the insulation monitoring device in these requirements is described as an isolation monitor, unlike in the IEC / EN standards as an insulation monitoring device. This is an American characteristic, which is a litlle confusing, but is accepted.

To sum up, protective and monitoring devices take on the same importance in electric vehicles and in charging stations as in other technical areas. Co-operation between the manufacturers of electric vehicles, operators of charging stations and the electro-technical industry is highly recommended and will allow to integrate the required measures into the electric vehicles. All prerequisites in terms of measurement techniques are given.

## Literature

[7.1] Danke, E.; Schütz, R.: Schutztechnik in Starkstromanlagen mit Nennspannungen bis 1.000 V im Bergbau. Schriftverkehr der Fa. Bender, Grünberg, 1991

[7.2] International convention for the safety of life at sea (1989): International maritime organization – London

[7.3] Lloyd's Register of Shipping – Rules and regulations for the classification of ships, Part 6: Control, Electrical, Refrigeration and Fire, Jan. 1984

[7.4] The Institution of Electrical Engineers: Recommendations for the electrical and electronic equipment of mobile and fixed offshore installations, 1983

[7.5] The Institution of Electrical Engineers: Regulations for the electrical and electronic equipment of ships, 1972

[7.6] Canadian Coast Guard – ship safety electrical standards: Coast guard ship safety branch, 1982 with amendment 1987

[7.7] Det Norske Veritas – Rules for classification of steel ships. Part 4 chapter 4: Electrical installations, Jan. 1990

[7.8] Det Norske Veritas – Rules for classification of steel ships. Part 5 chapter 3: Oil carriers, Jan. 1990

7.9] Det Norske Veritas – Rules for classification of mobile offshore units, Part 4 chapter 4: Electrical installations, Jan. 1991

[7.10] Bender, C.: Safe and reliable electrical power systems for navy ships. Firmenschrift Bender, Grünberg

[7.11] Harders, W.: DIN VDE 0129 Teil 201 – Eine deutsche Norm zur Elektrotechnik auf Schiffen. de Der Elektromeister & Deutsches Elektrohandwerk (1991) H. 13, H. 15 – 16

[7.12] Bauvorschrift für Schiffe der Bundeswehr – Marine – 30, E-Anlage. Koblenz: Bundesamt für Wehrtechnik und Beschaffung, 1986

[7.13] Bundesverband Solar-Mobil (BSM), URL: http://www.elektrofahrzeuge.net

[7.14] Coroller 96, Coroller, Patrick, ADEME, Frankreich. The low emission vehicle programs in France, 1996

[7.15] DAUG, Deutsche Automobilgesellschaft mbH: Erprobung von Elektrofahrzeugen der neuesten Generation auf der Insel Rügen, Abschlussbericht.

[7.16] Erlebnisse mit dem miniel. Ein Praxisbericht im Internet http://www.ralfwagner.de/mini.htm

[7.17] Naunin, Dietrich: Elektrische Straßenfahrzeuge, Technik, Entwicklungsstand und Einsatzgebiete. 2. Aufl., Expert-Verlag, 1994

[7.18] Underwriters Laboratories Inc. Standard for Safety: UL 2231-1, Personnel Protection Systems for Electric Vehicle (EV), General Requirements, Northbrook 2002

[7.19] Underwriters Laboratories Inc. Standard for Safety: UL 2231-2, Personnel Protection Systems for Electric Vehicle (EV), Specific Requirements for Protection Devices for Use in Charging Systems, Northbrook 2002

# 8 Insulation Resistance

The importance of insulation resistance in electrical systems, installations, devices and components was mentioned officially for the first time about one hundred years ago. Nowadays insulation resistance thresholds, their testing and monitoring, as well as indication and protective measures have become imperative. Residual current protective devices (RCDs), which trip at direct contact, applied in TT and TN systems are generally known. Less known however, is residual current monitoring as a measure for fault indication. In IT systems the absolute value of an installation is permanently monitored and an audible and/or visual warning signal initiated, if this value fall below a certain level.

The American NFPA (National Fire Protection Association) publishes statistics about fires and causes of fires in American households every year. Electrical distribution equipment fires ranked first in property damages in 1999 [8.1]. Electrical distribution equipment includes: fixed wiring; transformers or associated overcurrent or disconnect equipment; meters or meter boxes; power switch gear or overcurrent protection devices; switches, receptacles or outlets; light fixtures, lamp holders, light fixtures, signs, or ballasts; cords or plugs; and lamps or light bulbs.

In 1999, electrical distribution equipment in the home caused 40,100 structure fires, 226 civilian fire deaths, 1,166 civilian fire injuries, and $ 804.7 million in direct property damage.

During the five-year period from 1994 through 1998, electrical distribution equipment caused an annual average of 38,400 home structure fires, 352 civilian fire deaths, 1,343 civilian fire injuries, and $ 614.2 million in direct property damage.

Electrical distribution equipment fires ranked:

- fourth in number of home structure fires in 1999 and fifth during 1994-1998
- fifth in home fire deaths in 1999 and fourth during 1994-1998
- seventh in home fire injuries
- first in direct property damage in 1999 and second during 1994-1998

Electrical distribution equipment caused 11 – 13 % of the fires in one- and two-family dwellings or manufactured homes, but only 5 – 6 % of the fires in apartments. A study done by the U. S. Consumer Product Safety Commission in the mid 1980's examined detailed information about electrical equipment residential fires in specific cities. They found that improper alterations contributed to 37 % of the fires; improper initial installations factored in 20 % of the incidents; deterioration due to aging system components contributed to 17 % of the fires; improper use was a factor in 15 % of the incidents; inadequate electrical capacity contributed to another 15 %; faulty products were implicated in 11 %, and contributing factors were unknown in 6 % of the fires studied.

Other American safety research data (CFOI and SOII) shows that 2,287 U.S. workers died and 32.807 workers sustained days away from work due to electrical shock or electrical burn injuries between 1992 and 1998. The narrative, work activity, job title, source of injury, location and industry for each fatal electrical accident were examined. A primary causal factor was identified for each fatality. Electrical fatalities were categorized into five major groups. Overall, 44 % of electrical fatalities occurred in the construction industry. Contact with overhead power lines caused 41 % of all electrical fatalities. Electrical shock caused 99 % of fatal and 62 % of nonfatal electrical accidents. Comprising about 7 % of the U.S. workforce, construction workers sustain 44 % of electrical fatalities. Power line contact by mobile equipment occurs in many industries and should be the subject of focused research. Other problem areas are identified and opportunities for research are proposed. Improvements in electrical safety in one industry often have application in other industries. [8.2]

German research reports from 1990 on "causes of fatal accidents due to electric current in low voltage distribution systems" state that the cause was attributed to insulation faults, in numbers that are 26 out of 164 accidents in the group of operating errors [8.3]. In these cases the insulation had either been damaged or removed from devices by unskilled persons, or made ineffective through water.

Although these figures should not be overestimated, they do point out that preventive insulation measurements by TT and TN systems, as well as permanent insulation monitoring by IT systems is of great importance.

Let me quote at this point an abstract from the NFPA President & CEO, Mr Jim Shannon from the Spring 2003 edition of nfpa-journal in a comment he made on the deadly nightclub fires, two of which had occurred in Rhode Island and Chicago, which had tragically killed 118 young people and is titled: Nightclub tragedies underscore the need for safety improvements.

"Existing arrangements, retroactive application of code requirements, and inspections and permitting activities must also e critically examined.

NFPA has always rapidly incorporated lessons learned following significant fires throughout the last century. On countless occasions, our codes and standards have been the impetus for needed reforms nationally and worldwide.

The application and enforcement of those codes and standards have saved countless lives. But those responsible for updating codes and standards as well as the enforcement community must learn from and respond to these tragedies." [8.4]

When the insulation resistance is being considered, there are two aspects clearly distinguishable:

- the insulation resistance of de-energized systems without connected consumers
- the insulation resistance of energized systems with connected consumers

The significant differences in the insulation values of the different types of systems shall be clarified here since differences are often unrecognised, confused or unknown.

The dangers of applying electrical energy is known since the pioneering days. For good reason insurance engineers of the Phoenix Fire Office in London had demanded special safety measures for installing electrical wires by insulating them with a non-flammable insulation material and a double protective layer made of a solid durable substance.

## 8.1 History on Safety Regulations in Germany (1883)

Numerous fires in several industrial branches caused by electric current prompted German Fire Insurances to publish their first safety regulations for electrical installations on 20th August **1883**, basically dealing with arc and filament lamps. The publication of the German translation of "Regulations of the Phoenix Fire Office" for electric lighting and power stations in **1891**, a significant work at the time, created an awareness in the German public. The value of this work at the time was shown in 16 re-prints in the first eight years. For the first time it contained values on the insulation resistance, which was made proportionally dependent on the type of system and number of consumers connected.

German experts believed that making the insulation resistance dependent on the number of light bulbs was too inaccurate. The Association of Electro-Technique in Berlin (ETV) therefore published "Safety regulations for electrical high voltage installations against fire hazards" in December **1894**. §5 was on the insulation resistance.

The first comprehensive safety standards for electrical installations in Germany were published in January **1896** by the "Verband Deutscher Elektrotechniker e. V. (VDE)" (German Electro-technical Association), which addressed the insulation resistance in a whole chapter [8.5].

The most important prerequisite for avoiding personal and material damage by electrical equipment is an adequate insulation resistance. For that reason it is advisable to place a much higher demand on it, than necessary for operational reasons. Testing the insulation resistance is crucial for the evaluation of the safety-aspects of the installation.

## 8.2 Insulation Resistance – a Complex Matter

The insulation resistance against earth, respectively against the live conductor, respectively against exposed conductive parts of the live conductors in electrical installations or equipment are a highly complex matter, as they not only consist of just the insulation of the conductors, but also have clearances or contaminated and damp creepage distances. The most common representation is an equivalent circuit as shown in **Figure 8.1**, where a) represents insulation resistance independent of the voltage level, being in combination with b) which includes a voltage-dependent resistance, which is commonly called the insulation resistance. Low electrical voltages are considered in d). The leakage capacitance in c) is a relatively constant quantity, which only varies if the system is extended (cable length or added consumers).

Insulation faults, that are short circuits and earth faults are primary reasons for the safety considerations in electrical installations. Their share lies at 80 % to 90 % of the total number of systems faults.

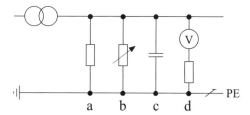

**Figure 8.1** Equivalent diagram for insulation resistances
a, b, c     constant and variable resistance and capacitive resistance
d         voltage with series resistance

## 8.3 Definitions

The International Electrotechnical Vocabulary defines insulation fault as follows:

Insulation fault: "An insulation fault is a defect in the insulation of an equipment, which can result either in abnormal current through this insulation or in a disruptive discharge" [IEV 604-02-02].

The standard series of IEC 61557, "Electrical safety in low voltage distribution systems up to 1000 V a.c. and 1500 V d.c. – Equipment for testing, measuring or monitoring of protective measures" offer a number of definitions, just to mention a few:

Part 6, "Residual current devices (RCD) in TT, TN and IT systems":

**Fault current:** "Current flowing to earth due to an insulation fault" [IEC 61557-6]

Part 8, "Insulation monitoring devices for IT systems":

**Insulation resistance:** "Resistance in the system being monitored, including the resistance of all the connected appliances to earth" [IEC 61557-8]

## 8.4 Influence Quantities

Usually, if an electrical installation or electrical equipment is brand-new, it can be assumed that the insulation resistance is in a good state. Naturally manufacturers of wires, motors, etc. are constantly improving the insulation condition of their installation for the practical application in the industry. But a number of factors have a deteriorating influence on the insulation and cause resistance, for example: mechanical damage, vibrations, excessive temperatures, dirt, oil, corrosion, moisture from industrial processes or from environmental and climatic conditions.

Every system has a specific insulation resistance against earth. In new installations this resistance is within the M$\Omega$ range. However, installations are affected by various external influences:

- electrical effects through overvoltage, overcurrent, frequency, lightning as well as magnetic and inductive influences
- mechanical effects due to shock/impact, flaw/bend, vibration and penetration of foreign bodies, such as nails
- environmental effects due to temperature/moisture, light/ultraviolet rays, chemical influence, pollution, animals

Furthermore every electrical installation and hence the electrical insulation are subject to aging effects, which also gradually deteriorate the insulation values.

Despite the greatest of care by the manufacturers of electrical equipment on the selection as well on the erection of electrical installations, it cannot be guaranteed that sooner or later even the best insulation material will deteriorate and cannot meet the requirements regarding electrical and mechanical stress. Consequences from resulting insulation faults may be:

- short circuits
- short circuits to earth
- short-circuit to exposed-conductive parts

In the case of a short-circuit to exposed-conductive parts, the conductive parts, which are not included in the operating circuit (inactive parts), are conductively connected to energized parts (active parts) of the operating circuit. At this process faults occur, which result in touch voltages. Even low fault currents, starting as creeping contact resistances may develop into arcs, short-circuits or short-circuits to earth. They may become dangerous, if in close proximity to flammable or combustible agents. In TN or TT systems this fire risk already comes into existence with the first insulation fault at the unearthed conductor, if the loss of electric energy at the fault location is too high.

In IT systems an electric energy loss of this magnitude is only possible, if insulation faults occur at two different phase conductors at the same time.

## 8.5 Insulation Measurement and Monitoring

Beside permanent monitoring of the complete systems and installations, testing the insulation resistance in de-energized systems is one of the most important safety tests for electrical installations and equipment.

### 8.5.1 Measurements in De-Energized Systems

The insulation resistance of de-energized systems or parts of an installation is measured with devices in accordance with IEC 61557-2 (see Chapter 17). There are certain requirements, these measuring devices must meet, which are likewise regulated by IEC standards. In this way the construction and test-engineers of an electrical installation may be able to achieve and compare proper measuring results. All protective measures, with or without protective conductor, require insulation measurement. Attention should be paid to de-energizing all test-objects before applying the insulation measurement methods between:

- all phase conductors and protective conductors
- neutral and protective conductors
- phase conductors
- phase and neutral conductors

Tests may also be conducted with energized equipment. If the insulation resistance of these consumers is too low, the devices have to be disconnected and installation and devices be tested separately.

### 8.5.2 Residual Current Monitoring in TN and TT Systems

TN and TT systems are a well tried and often applied measure of protection against indirect contact by disconnection via a residual current protective device (RCD). The idea behind this measure is to lead all conductors, which shall be monitored, except for the protective conductor of course, through a current transformer. In a faultless system the sum of all currents equal zero, this means that no voltage is induced in the current transformer. If a fault current flows to earth, the sum of all currents is unequal zero. The voltage generated in the secondary winding of the current transformer releases a trigger to activate the protective device for disconnection of the installation.

A less known method for indicating a fault, is the method of residual (or fault) current monitoring. IEC 62020, 3.3.1 defines a residual current monitor (RCM) as a "device or association of devices which monitors the residual current in an electrical installation, and which activates an alarm when the residual current exceeds the operating value of the device". RCMs are devices with an auxiliary source.

Although it is not an absolute measuring system, as is the case in IT systems, the RCMs are capable of detecting fault currents in TN and TT systems at an early stage. This allows appropriate steps to be taken without de-energizing the whole installation. A large setting range of the RCMs allows adaptation to special applications. However, insulation faults occurring symmetrically on all conductors against PE or earth are not detectable by residual current measurement.

### 8.5.3 Continuous Monitoring of the Absolute Insulation Value in IT Systems

The above description of the insulation resistance in principle applies to all types of distribution systems. The continuous monitoring of the absolute insulation value of the complete installation is only possible in IT systems with insulation monitoring. This means monitoring during operation and during the de-energized state of

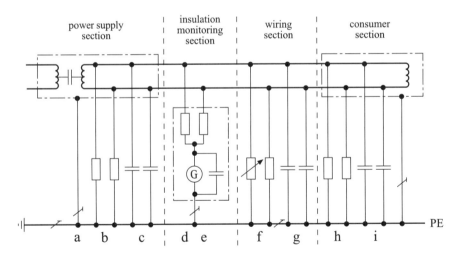

**Figure 8.2** Equivalent circuit diagram for a.c. IT system
a, b, c   transformer coupling capacitance, transformer leakage resistance, transformer leakage capacitance (capacitive resistance)
d, e   internal d. c. resistance and a. c. impedance of the insulation monitoring device
f   constant and varying insulation resistance
g   leakage capacitance of the installation (capacitive resistance)
h, i   insulation resistance and leakage capacitance (possibly also EMC filter capacitances) of connected current-using equipment

the installation. Monitoring devices are positioned between the active, unearthed IT system and earth, respectively protective conductor, which are continuously measuring and signalling visually or audibly, as soon as minimum or maximum set response values are reached.

IT systems are being supplied by a transformer, a generator, an accumulator or another independent current source. The special feature of these a.c. or d.c. systems is, that no live conductor of the system is directly earthed. The advantage is that the first short-circuit to exposed-conductive parts or to earth does not interfere with the operation of the installation. An electrical installation designed as an IT system consists of the power source, a wiring system and the connected electrical equipment. **Figure 8.2** is an example of an a.c. IT system represented as an equivalent circuit diagram. The absolute insulation level of energized IT systems of course is lower than the insulation level of individual sections as seen in Figure 8.2. In addition, influence quantities as described before are to be considered.

## 8.6  Complete Monitoring in IT Systems

Protective measures against indirect contact according to IEC 60364-4-41 are dependent on the coordination of the type of distribution system and the protective device. In IT systems with supplementary protective equipotential bonding, the system usually is monitored for first insulation faults by means of an insulation monitoring device (IMD), which consists of an IMD according to IEC 61557-8 with an alarm and test combination connected to the IMD.

These standards apply to monitoring devices which continuously monitor the insulation resistance to earth of unearthed IT a.c. systems, respectively IT a.c. systems with galvanically connected d.c. circuits with nominal voltages up to 1000 V, as well as IT d.c. systems. Earth fault relays using the asymmetry voltage (voltage shift) as the only measurement criterion in the event of an earth fault are not insulation monitoring devices in the interpretation of IEC 60364-4-41 or IEC 61557-8.

As required by various standards, insulation monitoring devices signal a decrease in the insulation resistance to earth below a minimum value. Attention must be paid to the fact that the insulation resistance in the IT system is the resistance of the system being monitored, including the resistances of all the connected equipment to earth. The system leakage capacitances also have to be taken into consideration when selecting and setting insulation monitoring devices. The latest insulation monitoring devices designed in accordance to this standard are also capable of signalling the failure of connected wires to the monitored system or initiating an alarm, if the protective conductor is disconnected.

The response times of the insulation monitoring devices are subject to the system leakage capacitances. In pure a.c. systems the response times usually are below 1 s, with a negligible system leakage capacitance of up to 1 μF. However, response times of up to 10 s (IEC 61557-8) in a.c. systems and in a.c. systems with d.c. components, response times of up to 100 s are permitted.

Taking into consideration that an IT system with supplementary protective equipotential bonding in the event of the first can be likened to the design of a TN system, it becomes obvious that longer response times do not have adverse effects on insulation monitoring devices. Longer measurement periods are also required in systems with very high leakage capacitances or voltage fluctuations.

## Literature

[8.1] NFPA Fire Analysis and Research, Quincy, MA, USA; U.S. Fire Problem Overview Report, 6/03, P. 70

[8.2] Cawley, J. C., Journal of safety research, Band 34 (2003) Heft 3, S. 241 – 248

[8.3] Zürnek, H.: Ursache tödlicher Stromunfälle bei Niederspannung. Schriftenreihe der Bundesanstalt für Arbeitsschutz-Forschung, Dortmund 1990, S. 30

[8.4] Jim Shannon, nfpa-Journal, The source for Fire, Electrical, Building – Life Safety Information, March/April 2003, www.nfpajournal.org, Volume 97, Number 2, U.S.A

[8.5] Rudolph, W.: VDE-Schriftenreihe, Bd. 39. Einführung in die DIN VDE 0100, Elektrische Anlagen von Gebäuden. 2. Aufl., Berlin · Offenbach: VDE VERLAG, 1999

# 9 Effects of Shock Current on Humans

The use of electrical equipment and devices exposes humans basically to two types of dangers [9.1]:
- the immediate or direct impact of electrical energy on a person in the event of a current path through the human body, and
- the indirect impact of damaging events caused by electrical energy, such as fires or switch-operations.

Electrical current flowing through the human body will **directly** negatively effect the body – the current itself is the danger. The effects on the body can range from a slight tingling sensation to severe conditions like muscular paralysis and ventricular fibrillation, which are the main causes of fatal electro accidents. The effects can also change the body substance in the form of coagulations and burns. This fact sets the dangers from electric current passing through the human body clearly apart from other dangers that come from electrical energy.

However electric current is only **indirectly** dangerous to the human body, when it comes from electric arcs or electro ophthalmia. In this case the dangers are heat and radiation, which can cause injuries like burns of any kind and severity. Usually this goes hand in hand with material damages, for instance in the event of fires caused by electric arcing. However, regardless of the initiating cause, one can say that these types of dangers are not specific to electrical current, since this kind of danger is also present in other types of technical installations. Still, the danger of the current flow is specific to using electrical equipment or operating electrical installations.

The current flow through the human body can be initiated through either:

- **direct contact** of conductive parts of electrical equipment or devices during normal operation under voltage. This precipitates that the protective measure of insulation, barrier, enclosure, obstacle or placing out of reach, which are normally placed between person and exposed-conductive part, has been removed or damaged; or
- **indirect contact** of electrical equipment or devices during normal operation not voltage-free, but through fault conditions energized. Contact with such parts, so-called solid body, is normally unavoidable when operating electrical equipment.

If direct or indirect contact occurred, the extend of the impact on the human body is determined by a number of factors:
- the current path,

- the touch voltage
- the duration of the current flow
- the frequency,
- the different resistances of the human body (e.g. degree of moisture on the skin)
- the surface area of contact,
- the pressure exerted and
- the temperature

Summarizing it can be said that the severity of impact of the current flow on a person is dependent on the intensity of the touch voltage, the value of the total body impedance (body and skin) and the duration the person was exposed to the current.

## 9.1 The Effects of Current on Human Beings and Livestock

In the late 1960s an IEC working group began work on the IEC Technical Report about the effect of electrical current on humans. They first published their findings in 1974. The practical applications of the report however resulted in various difficulties. The findings on impedances (high-resistance fault on exposed-conductive-parts) were so incomplete, that new research was needed. Due to the inherent danger of experimenting with electrical energy, it was unknown, which voltages a living human body was capable of withstanding. Experimental results, available from measurements carried out principally on corpses or from animal research, only allowed rough estimations of the effects on the living organism. This has led the Austrian scientist Professor Biegelmeier in 1976 to conduct experiments with touch voltages up to 200 V on his own body. His sensational experiments and further research and measurements have formed the basis to experimental electropathology and led to the revision of Technical Report 479.

Revision on the IEC 479-report was completed in 1984. But even the second edition of this report still had gaps. Later research work conducted on other physical accident parameters, especially the waveform and frequency of the current and the impedance of the human body, led to more understandings. The third revision in 1994 was therefore considered necessary and can be viewed as the logical development and evolution of IEC 479, a concept with four parts.[1]

In 2005 the latest revised edition of the Technical Specifications IEC 60479-1, Ed. 4.0 was published. It provides basic guidance on the effects of shock current on human beings and livestock, for use in the establishment of electrical safety requirements.

---

[1] See Chapter 17 for listing of this series

The form of the document as has been adopted summarizes results so far achieved which are being used by Technical Committee 64 as a basis for fixing requirements for protection against electric shock in electrical installations. Maintenance Team 4 has the special task of working on a standard regarding the „Effects of current passing through the body". The results are considered important enough to justify an IEC publication which may serve as a guide to other IEC committees and countries having need of such information.

This Technical Specification applies to the thresholds of ventricular fibrillation which is the main cause of death by electric current. The analysis of results of recent research work on cardiac physiology and on the fibrillation threshold, alltogether, has made it possible to better appreciate the influence of the main physical parameters and, especially, the duration of the current flow.

Publication IEC 60479 contains information about body impedance and body current thresholds for various physiological effects. This information can be combined to derive estimates of a.c. and d.c. touch voltage thresholds for certain body current pathways, contact moisture conditions, and skin contact areas. Information about touch voltage thresholds for physiological effects is contained in the IEC publication IEC 61201 (Chapter 17).

The following clauses endeavour to describe the aspects relevant to this book.

### 9.1.1  Scope and Object of IEC 60479-1:2005-07

Score and object of IEC 60479-1 are as follows:

„For a given current path through the human body, the danger to persons depends mainly on the magnitude and duration of the current flow. However, the time/current zones specified in the following clauses are, in many cases, not directly applicable in practice for designing measures of protection against electric shock. The necessary criterion is the admissible limit of touch voltage (i.e. the product of the current through the body called touch current and the body impedance) as a function of time. The relationship between current and voltage is not linear because the impedance of the human body varies with the touch voltage. Data on this relationship is therefore required. The different parts of the human body (such as the skin, blood, muscles, other tissues and joints) present to the electric current a certain impedance composed of resistive and capacitive components.

The values of body impedance depend on a number of factors, in particular, on current path, on touch voltage, duration of current flow, frequency, degree of moisture of the skin, surface area of contact, pressure exerted and temperature. The impedance values indicated in the Technical Specification result from a close examination of the experimental results available from measurements carried out principally on corpses and on some living persons."

Knowledge of the effects of alternating current is primarily based on the findings related to the effects of current at frequencies of 50 Hz or 60 Hz, which are the most common in electrical installations. The values given are, however, deemed applicable over the frequency range from 15 Hz to 100 Hz, threshold values at the limits of this range being higher than those at 50 Hz or 60 Hz. Principally the risk of ventricular fibrillation is considered to be the main mechanism of death of fatal electrical accidents.

Accidents with direct current are much less frequent than would be expected from the number of d.c. applications, and fatal electrical accidents occur only under very unfavourable conditions, for example, in mines. This is partly due to the fact that with direct current, the let-go of parts gripped is less difficult and that for shock durations longer than the period of the cardiac cycle, the threshold of ventricular fibrillation is considerably higher than for alternating current.

### 9.1.2 Electrical Impedance of the Human Body

The values of the internal body impedance depend on a number of factors and, in particular, on the current path, on touch voltage, duration of current flow, frequency, degree of moisture of the skin, surface area of contact, pressure exerted and temperature.

A schematic diagram for the impedance of the human body is shown in **Figure 9.1**.

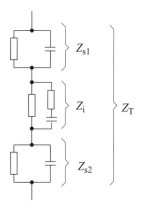

**Figure 9.1** Impedances of the human body (IEC 60479-1, Figure 1)
$Z_i$ Internal impedance
$Z_{s1}$; $Z_{s2}$ Impedance of the skin
$Z_T$ Total impedance

## 9.1.3 Sinusoidal alternating Current 50/60 Hz for Large Surface Areas of Contact

The value of the total body impedance in **Table 9.1** is valid for living human beings and a current path hand to hand for large surface areas of contact (order of magnitude 10 000 mm$^2$) in dry conditions.

The range of the total body impedances for touch voltages up to 700 V for large surface areas of contact in dry, water-wet and saltwater-wet conditions for a percentile rank of 50 % of the population is presented in Table 9.1.

The values for Table 9.1 represent the best knowledge on the total body impedances $Z_T$ for living adults. On the knowledge at present available the total body impedance $Z_T$ for children is expected to be somewhat higher but of the same order of magnitude.

| Touch voltage V | Values for the total body impedance $Z_T$ ($\Omega$) that are not exceeded for a percentage (percentile rank) of | | |
|---|---|---|---|
| | 5 % of the population | 50 % of the population | 95 % of the population |
| 25 | 1750 | 3250 | 6100 |
| 50 | 1375 | 2500 | 4600 |
| 75 | 1125 | 2000 | 3600 |
| 100 | 990 | 1725 | 3125 |
| 125 | 900 | 1550 | 2675 |
| 150 | 850 | 1400 | 2350 |
| 175 | 825 | 1325 | 2175 |
| 200 | 800 | 1275 | 2050 |
| 225 | 775 | 1225 | 1900 |
| 400 | 700 | 950 | 1275 |
| 500 | 625 | 850 | 1150 |
| 700 | 575 | 775 | 1050 |
| 1000 | 575 | 775 | 1050 |
| Asymptotic value = Internal impedance | 575 | 775 | 1050 |

**Table 9.1** Total body impedance $Z_T$ for a current path hand to hand a. c. 50/60 Hz, for large surface areas of contact in dry condition (IEC 60479-1, Table 1)

### 9.1.4 Threshold of Ventricular Fibrillation

The threshold of ventricular fibrillation depends on physiological parameters (anatomy of the body, state of cardiac function, etc.) as well as on electrical parameters (duration and pathway of current flow, current characteristic, etc.). A description of heart activity is given in **Figures 9.2 and 9.3**.

With sinusoidal a.c. (50 Hz or 60 Hz) there is a considerable decrease in the threshold of fibrillation if the current flow is prolonged beyond one cardiac cycle. This effect results from the increase in inhomogeneity of the excitatory state of the heart due to the current induced extra-systoles.

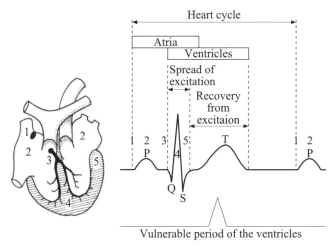

**Figure 9.2** Occurrence of the vulnerable period of ventricles during the cardiac cycle. The numbers designate the subsequent stages of propagation of the excitation (from IEC 60479-1, Figure 17)

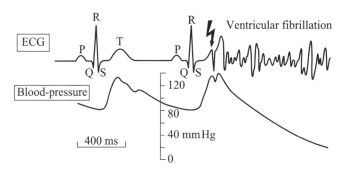

**Figure 9.3** Triggering of ventricular fibrillation in the vulnerable period. Effects on electrocardiogram (ECG) and blood-pressure (from IEC 60479-1, Figure 18)

For shock durations below 0.1 s, fibrillation may occur for current magnitudes above 500 mA, and is likely to occur for current magnitudes in the order of several amperes, only if the shock falls within the vulnerable period. For shocks of such intensities and durations longer than one cardiac cycle reversible cardiac arrest may be caused.

For duration of current flow longer than one heart period **Figure 9.4** shows a comparison between the thresholds of ventricular fibrillation from animal experiments and for human beings calculated from statistics of electrical accidents.

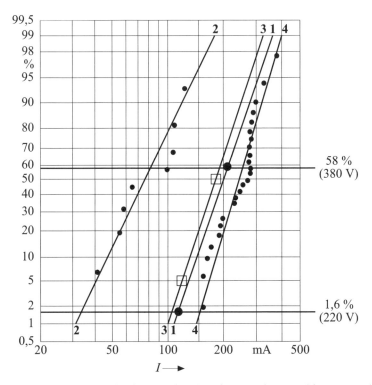

**Figure 9.4** Fibrillation data for dogs, pigs, sheep from experiments and for persons calculated from statistics of electrical accidents with transversal direction of current flow "hand – hand" and touch voltages $U_T = 220$ V and 380 V a. c. with body impedances $Z_T$ (5 %) (IEC 60479-1, Figure 19)

1     fibrillation data for persons calculated from statistics of accidents ($U_T = 220$ V, 1.6 %, $U_T = 380$ V, 58 %)
2     fibrillation data for dogs, duration of current flow 5 s
3     fibrillation data for pigs, duration of current flow $t > 1.5 \times$ heart-period
4     fibrillation data for sheep, duration of current flow 3 s
(1)   values corrected with the heart-current factor $F = 0.4$

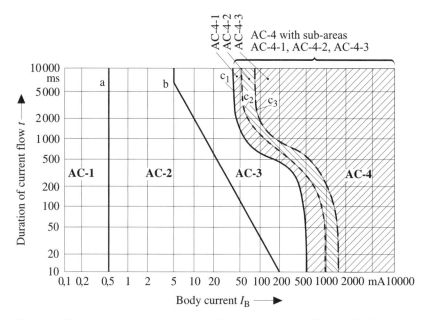

**Figure 9.5** Conventional time/current zones of effects of a.c. currents (15 Hz to 100 Hz) on persons for a current path corresponding to left hand to feet (for explanation see **Table 9.2**)
(IEC 60479-1, Figure 20)

In adapting the results from animal experiments to human beings, an empirical curve c1 (see **Figure 9.5**) was conventionally established for a current path left hand to both feet, below which fibrillation is unlikely to occur. The high level for short durations of exposure between 10 ms and 100 ms was chosen as a descending line from 500 mA to 400 mA. On the basis of information on electrical accidents, the lower level for durations longer than 1 s was chosen as a descending line from 50 mA at 1s to 40 mA for durations longer than 3s. Both levels were connected by smooth curve.

By statistical evaluation of animal experiments, curve c2 and curve c3 (see Figure 9.5) have been established defining a probability of fibrillation of about 5 % and 50 % respectively.

Curves c1, c2 and c3 apply for current path left hand to both feet.

### 9.1.5 Description of Time/Current Zones for a.c. 15 Hz to 100 Hz

**Table 9.2** explains the time/current zones for the hand to feet pathway (see Figure 9.5).

| Zones | Boundaries | Physiological effects |
|---|---|---|
| AC-1 | up to 0,5 mA curve a | Perception possible but usually no startle reaction |
| AC-2 | 0,5 mA up to curve b | Perception and involuntary muscular contractions likely but usually no harmful electrical physiological effects |
| AC-3 | Curve b and above | Strong involuntary muscular contractions. Difficulty in breathing. Reversible disturbances of heart function. Immobilisation may occur. Effects increasing with current magnitude. Usually no organic damage to be expected. |
| AC-4[1] | Above curve $c_1$ | Pathophysiological effects may occur such as cardiac arrest, breathing arrest, and burns or other cellular damage. Probability of ventricular fibrillation increasing with current magnitude and time. |
|  | $c_1 - c_2$ | AC-4.1 Probability of ventricular fibrillation increasing up to about 5 %. |
|  | $c_2 - c_3$ | AC-4.2 Probability of ventricular fibrillation up to about 50 %. |
|  | beyond $c_3$ | AC-4.3 Probability of ventricular fibrillation above 50 %. |

[1] For durations of current flow below 200 ms ventricular fibrillation is only initiated within the vulnerable period if the relevant thresholds are surpassed. As regards ventricular fibrillation this figure relates to the effects of current which flows in the path left hand to feet. For other current paths the heart current factor has to be considered.

**Table 9.2** Time/current zones for a.c. 15 Hz to 100 Hz for hand to feet pathway: Summary of zones of Figure 9.5 (from IEC 60479-1, Table 11)

## 9.2 Electro-pathological realizations

In addition to the Technical Specifications IEC 60479 some ideas from Gottfried Biegelmeier's work [9.2] are worth mentioning:

- The internal body impedance in a current range of 5 V to 5000 V a.c. is independent of the value of the applied voltage.
- In the area of broken skin the total impedance sinks with rising voltage and current density.
- The body impedance for a.c. current is equal to that of d.c. current. The total impedance of the human body with d.c. touch voltage under 100 V is above the d.c. current, because of the blocking effects of the body-capacitances.
- Essentially the internal body impedance is a pure equivalent resistance.

- The total impedance of the human body before and during breaking of the skin is strongly dependent on the moisture content at the breaking point of the skin. The moisture content is however irrelevant on the internal body impedance.
- At voltages above approximately 200 V the skin has no protection any more, and the contact surface is irrelevant to the outcome of the accident. The same applies to the state of the skin in relation to moisture and temperature.
- The current path through the human body in relation to its body impedance is relevant for the outcome of the accidents, at electro accidents and touch voltages above 100 V. Moisture and contact surface have little relevance.
- At touch voltages below approximately 100 V the body impedance drops while the frequency is rising. Is the contact surface small and the skin dry, the body impedance is in reverse proportion. Above approximately 1000 Hz the body impedance is coming closer to the values of the internal body impedance for the respective current path through the human body.
- d.c. current is less dangerous than a.c. current of 50/60 Hz because of the lack of the let-go threshold and the higher fibrillation threshold, during the current flow through the human body over a cardiac cycle.
- a.c. currents over 1000 Hz are less dangerous than the frequencies 50/60 Hz, which are normally used. At frequencies over 10.000 Hz usually neither muscular contractions occur nor is fibrillation to be expected.

## 9.3  Protective Measures Against Shock Current

Protective measures against dangerous shock currents are described in IEC 60364-4-41, see chapter 4 for a more detailed explanation.

## 9.4  Accidents Involving Electrical Current

Statistically accidents involving electrical current are way down the list, despite their dangerous potential. This can be attributed to the stringent regulations of national and international standards. Fact is that there are still a great number of fatal accidents, which puts the responsibility on to those making decisions on safety for others, to adhere to safety requirements and to make sure, that the protective measures adopted, are effective.

German statistics [9.3] gathered between 1970 and 2001 on fatal accidents involving electric current in private homes and in the working environment, as seen in **Table 9.3**, show a decreasing trend. This is quite remarkable, since the number of private homes and the production levels of electrical operating equipment has been steadily increasing. The number of accidents yearly recorded were decreasing from

| Year | Accidents through electrical cablewire systems and devices | | Miscellaneous accidents | Non-describe accidents | Total |
|---|---|---|---|---|---|
| | in private households | in industry and commerce | | | |
| 1970 | 87 | 69 | 44 | 56 | 256 |
| 1971 | 71 | 81 | 35 | 65 | 252 |
| 1972 | 88 | 66 | 39 | 73 | 266 |
| 1973 | 76 | 60 | 54 | 80 | 270 |
| 1974 | 87 | 55 | 29 | 64 | 235 |
| 1975 | 87 | 46 | 34 | 54 | 221 |
| 1976 | 82 | 46 | 44 | 33 | 205 |
| 1977 | 73 | 34 | 34 | 36 | 177 |
| 1978 | 74 | 39 | 26 | 32 | 171 |
| 1979 | 48 | 30 | 24 | 52 | 154 |
| 1980 | 61 | 45 | 18 | 42 | 166 |
| 1981 | 50 | 25 | 20 | 55 | 150 |
| 1982 | 56 | 31 | 22 | 48 | 157 |
| 1983 | 58 | 31 | 20 | 52 | 161 |
| 1984 | 49 | 27 | 14 | 37 | 127 |
| 1985 | 39 | 24 | 12 | 34 | 109 |
| 1986 | 48 | 14 | 17 | 39 | 118 |
| 1987 | 39 | 17 | 7 | 25 | 88 |
| 1988 | 32 | 22 | 11 | 34 | 99 |
| 1989 | 42 | 24 | 8 | 47 | 121 |
| 1990 | 30 | 31 | 8 | 77 | 148 |
| 1991 | 23 | 35 | 10 | 40 | 108 |
| 1992 | 49 | 30 | 22 | 51 | 152 |
| 1993 | 31 | 32 | 12 | 38 | 113 |
| 1994 | 29 | 30 | 20 | 31 | 110 |
| 1995 | 27 | 31 | 20 | 16 | 94 |
| 1996 | 33 | 31 | 5 | 32 | 101 |
| 1997 | 26 | 25 | 12 | 29 | 92 |
| 1998 | 32 | 18 | 17 | 21 | 88 |
| 1999 | 28 | 16 | 17 | 25 | 86 |
| 2000 | – | – | – | – | 100 |
| 2001 | 28 | 6 | 8 | 24 | 66 |
| 2002 | * | * | * | * | 65 |
| 2003 | 5 | 3 | 15 | 37 | 67 |
| 2004 | ** | ** | ** | ** | 49 |

\* Restructuring of the statistical data
\*\* No detailed data available

**Table 9.3** Fatal accidents caused by electric current (statistic from the Statistisches Bundesamt, Wiesbaden, Germany).
Note: Since 2000 the statistic on accident locations has been revised, so only the total number of electro accidents is given for that year.

319 in 1954 to just 99 in 2004. However since 1988 the number of fatal electro accidents were rising again and falling to just 49 in 2004.

Observing the total number of electro accidents in relation to the accident location, it is suggested that the decrease was decisively influenced by a decrease in work-related accidents involving electrical current in the industrial and commercial sectors.

Different statistical material allows the conclusion that fatal work related accidents in Germany have an altogether decreasing trend, undoubtedly due to the fact that the industry is making a great effort to implement safety measures recommended by specialized people and institutions. These efforts which were mainly initiated by the Employers Liability Insurance Association (Germany) and other renown European institutions also include further development of safety equipment, testing of electrical installations and electrical operating equipment, as well as specific information on safety and protection by specially skilled electro-technical persons.

## Literature

[9.1] Edwin, K.W.; Jakli, G.; Thielen, H.: Zuverlässigkeitsuntersuchungen an Schutzmaßnahmen in Niederspannungsverbraucheranlagen. Bundesanstalt für Arbeitsschutz und Unfallforschung, Dortmund. Forschungsbericht Nr. 221, 1979, S. 4

[9.2] Biegelmeier, G.: Wirkung des elektrischen Stromes auf Menschen und Nutztiere. Lehrbuch der Elektropathologie. Berlin u. Offenbach: VDE VERLAG, 1986

[9.3] Kieback, D.: Steigende Tendenz bei Selbstmorden mit Strom, etz Elektrotech. Z. 111 (1990) H. 9

# 10 International Standards for Insulation Monitoring Devices

There are a number of International standards in existence for insulation monitoring systems. These systems include insulation monitoring devices, remote alarm indicator and test combinations as well as insulation fault location systems. The following chapter describes these standards, provides information about the field of application and some technical data.

## 10.1 Insulation Monitoring Devices (IMD) for Monitoring a. c. Systems in Accordance with IEC 61557-8

The International series of standards IEC 61557 deals with the electrical safety in low voltage distribution systems up to 1000 V a. c. and 1500 V d. c. and describes equipment for testing, measuring or monitoring of protective measures. The increasing demand in the 1980s for galvanically connected d. c. circuits in industrial a. c. systems has made it necessary to develop part 8 of the IEC 61557-series.

IEC 61557-8 is sub-titled "Insulation monitoring devices for IT systems". The scope is as follows:

"This part of IEC 61557 specifies the requirements for insulation monitoring devices which permanently monitor the insulation resistance to earth of unearthed IT a. c. systems, for IT a. c. systems with galvanically connected d. c. circuits having nominal voltages up to 1 000 V a. c., as well as of unearthed IT d. c. systems with voltages up to 1 500 V d. c. independent from the method of measuring.

*Note 1: IT systems are described in IEC 60364-4-41 amongst other literature. Additional data for a selection of devices in other standards should be noted.*
*Note 2: Various standards specify the use of insulation monitoring devices in IT systems. In such cases, the objective of the equipment is to signal a drop in insulation resistance below a minimum limit.*
*Note 3: Insulation monitoring devices according to this part of IEC 61557 may also be used for de-energized electrical systems."*

The following requirements shall apply:

"Insulation monitoring devices shall be capable of monitoring the insulation resistance of IT systems including symmetrical and asymmetrical components and to

| Marking | Pure a.c. systems | AC systems with galvanically connected d.c. circuits and d.c. systems |
|---|---|---|
| Response time $t_{an}^a$ | $\leq 10$ s at $0{,}5 \cdot R_{an}$ and $C_e = 1$ µF | $\leq 100$ s at $0{,}5 \cdot R_{an}$ and $C_e = 1$ mF |
| Permanently admissible extraneous d.c. voltage $U_{fg}$ | According to the indications of the manufacturer | $\leq$ peak value $1{,}15 \cdot U_n$, not applicable for d.c. systems |

| Marking | For all systems |
|---|---|
| Peak value of the measuring voltage $U_m$ | At $1{,}1 \cdot U_n$ and $1{,}1 \cdot U_S$ as well as $R_F = \infty$ : $\leq 120$ V |
| Measuring current $I_m$ | $\leq 10$ mA at $R_F = 0$ |
| Internal impedance $Z_i$ | $\geq 30$ $\Omega$/V rated system voltage, at least $\geq 15$ k$\Omega$ |
| Internal resistance $R_i$ | $\geq 30$ $\Omega$/V rated system voltage, at least $\geq 1{,}8$ k$\Omega$ |
| Permanently admissible nominal voltage | $\leq 1{,}15 \cdot U_n$ |
| Relative (percentage) uncertainty [b] | $\pm 15$ % the rated response value $R_F$ |
| Climatic environmental conditions | Operation: [c] class 3K5 (IEC 60721-3-3), $-5$ °C to $+45$ °C<br>Transport: class 2K3 (IEC 60721-3-2), $-25$ °C to $+70$ °C<br>Storage: class 1K4 (IEC 60721-3-1), $-25$ °C to $+55$ °C |

[a] In IT systems, where the voltage is altered at low speed (e.g. converter systems with low speed control procedures or d.c. motors with low speed variation), the response time depends on the lowest operational frequency between the IT system and earth. These response times may differ from the above-defined response times.

[b] The relative uncertainty is defined with the following reference conditions:
• temperature: $-5$ °C and $+45$ °C;
• voltage: 0 % and 115 % of nominal voltage output of 85 % and 110 % of the rated supply voltage;
• frequency: rated frequency;
• leakage capacitance: 1 µF.
If the response value is adjustable, the range of response values which are not in the specified limits shall be marked for example. by dots at the limits of the range or the ranges. Information about the relative uncertainty within the working range specified by the manufacturer, but for leakage capacitances above 1 µF for frequencies below or above the nominal frequency or frequency range, shall be included in the documentation.

[c] Except: condensation and formation of ice.

**Table 10.1** Requirements on insulation monitoring devices according to IEC 61557-8, Ed. 2.0:2006, Table 1

give a warning if the insulation resistance between the system and earth falls below a predetermined level.

*Note 1: A symmetrical insulation deterioration occurs when the insulation resistance of all conductors in the system to be monitored decreases (approximately) similarly. An asymmetrical insulation deterioration occurs when the insulation resistance of, for example, one conductor decreases substantially more than that of the other conductor(s).*

*Note 2: So-called earth fault relays using a voltage asymmetry (voltage shift) in the presence of an earth fault as the only measurement criterion, are not insulation monitoring devices in the interpretation of this part of IEC 61557.*

*Note 3: A combination of several measurement methods, including asymmetry monitoring, may become necessary for fulfilling the task of monitoring under special conditions on the system."*

**Table 10.1** gives some important technical data on the requirements applicable to insulation monitoring devices according to IEC 61557-8.

## 10.2 Insulation Monitoring Devices (IMD) in Accordance with IEC 60364-5-53

The International standard for insulation monitoring devices is IEC 60364-5-53, "Electrical installation of buildings – Part 5-53: Selection and erection of electrical equipment – Isolation, switching and control". Its 3rd edition was published in June 2002. This standard is currently under review, edition 4.0. has the status of a Committee Draft for Vote (CDV) and is numbered IEC 64/1546/RVC (Result of Voting on CDV):2006-06. As the whole series IEC 60364 is being completely restructured, so are title and structure of IEC 60364-5-53. Clause 538 is about "Monitoring devices". Because of their relevance to this book, subclauses 538.1 to 538.3 are quoted here.

Clause 538 clarifies the term "monitoring" with a definition from IEV 351-15-24, as "a function intended to observe the operation of a system or part of a system to verify correct function or detect incorrect functioning by measuring system variables and comparing the measured values with specified values".

**538.1 Insulation monitoring devices for IT systems (IMDs)**

An IMD is intended to be permanently connected to an IT system and to continuously monitor the insulation resistance of the complete system (secondary side of the power supply and the complete installation supplied by this power supply) to which it is connected.

*Note: An IMD is not intended to provide protection against electric shock.*

538.1.1 In accordance with the requirements of 411.6.3.1 of part 4-41[1], an IMD shall be installed in IT systems. The IMD shall be in accordance with IEC 61557-8, unless they comply with 538.1.4.

Instructions shall be provided indicating that when the IMD detects a fault to earth, the fault shall be located and eliminated, in order to restore normal operating conditions as soon as possible.

### 538.1.2 Installation of insulation monitoring devices (IMDs)

The "line" terminal(s) of the IMD shall be connected as close as practicable to the origin of the system either to:

- the neutral point of the power supply, or
- an artificial neutral point with impedances connected to the line conductors, or
- a line conductor or several line conductors.

Where in a multi-phase system, the IMD is connected between one phase and earth, it shall be suitable to withstand at least the phase-to-phase voltage between its "line" terminal and its "earth" terminal.

*Note: This voltage appears across these two terminals in case of a single insulation fault on another phase conductor.*

For d.c. installations, the "line" terminal(s) of the IMD shall be connected either directly to the mid-point, if any, or to one or all of the supply conductors.

The "earth" or "functional earth" terminal of the IMD shall be connected to the main earthing terminal of the installation.

The supply circuit of IMD shall be connected either to the installation on the same circuit of the connecting point of the "line" terminal and as close as possible to the origin of the system, or to an auxiliary supply.

The connecting point to the installation shall be selected in such a way that IMD is able to monitor the insulation of the installation in all operating conditions.

Where the installation is supplied from more than one power supply, connected in parallel, one IMD per supply shall be used, provided they are interlocked in such a way that only one IMD remains connected to the system. All other IMDs monitor the disconnected power supply enabling the reconnection of this supply without any pre-existing insulation fault.

### 538.1.3 Adjustment of the insulation monitoring device (IMD)

IMD is designed for signalling any important reduction of the insulation level of the system in order to find the cause before a second insulation fault occurs, thus avoiding any power supply interruption.

---

[1] See Chapter 4

Consequently, the IMD shall be set to a lower value corresponding the normal insulation of the system when operating normally with the maximum of loads connected.

IMDs, installed in locations, where persons other than instructed (BA4) or skilled (BA5) persons have access to their use, shall be designed or installed in such a way, that it shall be impossible to modify the settings, except by the use of a key, a tool or a password.

### 538.1.4 Passive insulation monitoring devices

In some particular d. c. IT installations, including only two conductors, passive IMD, without the injection of current or voltage into the system, may be used, provided that the insulation of all live distributed conductors is monitored. All exposed conductive parts of the installation are interconnected circuit conductors are selected and installed so as to reduce the risk of earth fault to a minimum circuit conductors are selected and installed as such as to reduce the risk of earth fault.

### 538.2 Equipment for insulation fault location in IT systems

Equipment for insulation fault location shall be in accordance with IEC 61557-9. Where an IT system has been selected for continuity of service, it is recommended to combine the IMD with devices enabling the fault location on load. Their function is to indicate the faulty circuit when the insulation monitoring device has detected an insulation fault.

### 538.3 Monitoring off-line circuits

The monitoring of off-line circuits may be performed with IMD in TN-S, TT and IT systems provided that all live conductors of the monitored circuits are disconnected.

*Note: The above arrangement is intended to ensure that the safety equipment is allowed to work without intervention of supply during the emergency.*

The reduction of the insulation level shall be indicated locally by a visual or an audible signal with the choice of remote indication.

The IMD shall be connected between earth and a live conductor of the monitored equipment. The measuring circuit shall be automatically disconnected when the equipment is energized.

The IMD may be used for this purpose in all types of system earthing, except in TN-C systems.

## 10.3 Insulation Monitoring Devices (IMD) in accordance with the American ASTM Standards [10.1]

### 10.3.1 ASTM F 1207M-96:2002, Standard Specification for Electrical Insulation Monitors for Monitoring Ground Resistance in Active Electrical Systems

"This specification covers electrical insulation monitoring devices, intended as permanently installed units, for use in the detection of ohmic insulation faults to ground in active, ac ungrounded electrical systems."

It is "not intended to cover devices which are not intended for operation for: dc ungrounded systems or ac ungrounded systems with dc components unless ac to dc conversion is isolated from the monitored system with transformers."

### 10.3.2 ASTM F 1669M-96: 2002, Standard Specifications for Insulation Monitors for Shipboard Electrical Systems

"This specification covers two types of electrical systems insulation monitoring devices.
1. Type I is an ac device intended as a permanently installed unit for use in the detection of ohmic insulation faults to ground in active ac ungrounded electrical systems up to 1000 VAC, having dc components up to 1500 VDC.
2. Type II is a dc device intended as a permanently installed unit for use in the detection of ohmic insulation faults to ground in dc ungrounded electrical systems up to 1500 VDC.

This specification does not cover devices that are intended for operation in ac ungrounded systems without dc components."

### 10.3.3 ASTM F 1134-94:2002, Standard Specifications for Insulation Resistance Monitor for Shipboard Electrical Motors and Generators

"This specification covers monitoring devices (monitors) for the automatic detection and signalling of low insulation resistance values in idle electrical motors or generators, or both.
1. Monitors are intended for permanent installation in both existing or new panels and controller enclosures designed for marine application.
2. The values stated in SI units are to be regarded as the standard. The values given in parentheses are for information only.
3. The following safety hazards caveat pertains only to the test method described in this specification: This standard does not purport to address all of the safety concerns, if any, associated with its use. It is the responsibility of the user of this standard to establish appropriate safety and health practices and determine the applicability of regulatory limitations prior to use."

## 10.4 Difference between Insulation Monitoring Devices and Residual Current Monitors in Accordance with IEC 62020

In Pretoria in 1991 the IEC (International Electrotechnical Commission) came to the decision to create a standard for insulation monitoring devices. The application for this standard was based on the Norwegian document 23E (Norway) 18, which had been prepared by IEC SC 23E, Working Group 6. For the first time a distinction was made between "active" insulation monitoring and "passive" residual current monitoring.

The task was given to SC 23E, WG6 to define a standard for passive monitoring devices, based on the measuring principle of residual current measurement. As is known, asymmetrical insulation faults in electrical systems lead to residual currents.

Creating a standard for active monitoring devices was a task given earlier to IEC TC85 Working Group 8.

It was decided in SC23E WG 6, that a passive monitoring device would be termed "Residual Current Monitor (RCM)". The International Standard for RCMs is IEC 62020, Electrical accessories – Residual current monitors for household and similar uses (RCMs). The standard for IMDs is IEC 60364-5-53 (see 10.2).

The distinctive features between IMD and RCM is constituted in the scope of IEC 62020 as a note:

*Note: A RCM is distinguished from an IMD in that it is passive in its monitoring function and only responds to an unbalanced fault current in the installation being monitored.*

*An IMD is active in its monitoring and measuring functions in that it can measure the balanced insulation resistance or impedance in the installation (see IEC 61557-8)."*

## 10.5 Equipment for Insulation Fault Location in IT Systems

IEC TC85 WG8 was commissioned to work out another part for standard series IEC 61557. In September 1999 IEC 61557-9 was published, sub-titled: Equipment for insulation fault location in IT systems.

This part of the standard specifies the requirements for insulation fault location systems, which are capable of localizing symmerical as well as asymmetrical insulation deteriorations in any part of unearthed IT a.c. systems and unearthed IT a.c. systems with galvanically connected d.c. circuits, having nominal voltages of up to 1000 V a.c., as well as d.c. systems up to 1500 V, independent of the measuring principle. This part of IEC 61557 only applies in conjunction with part 1 of the series.

Note 1: IT systems are also described in IEC 60364-4-41. If other standards require additional specifications for the selection of devices, they have to be considered as well.

Note 2: An insulation fault location system usually consists of several modules (see **Figure 10.1**). All or several modules may be combined into one unit. The modules and their respective functions are briefly described as follows:

- Insulation monitoring devices (IMD) according to IEC 61557-8
- Control device, fixed or portable: the control device establishes the sequence of the test procedures and contains logical links for the location of insulation faults and fault indication
- Test device, fixed or portable: the test device contains the test current generator
- Coupling unit, fixed or portable: the coupling unit and the test device together establish the connection to the circuit, to be monitored
- Residual current transformer or current clamp: residual current transformer are used for detecting the test current; are connected to an evaluation unit
- Evaluation device, fixed or portable: current transformer are connected to the evaluation device for detecting the test current

Note 3: In note 2 (Figure 10.1) examples are listed for the modules of insulation monitoring devices. The fact that equipment may be fixed or portable might lead to differences in the module configuration.

Note 4: Insulation fault location systems with independent test current source may be used for insulation fault location in de-energized circuits.

The following is a definition for insulation fault location system according to IEC 61557-9:

*"Usually an insulation monitoring device which permanently monitors the insulation resistance of an unearthed IT system and additional insulation fault location modules, which are activated if an insulation fault is detected. The insulation fault location modules simultaneously may perform the function of injecting a test current between the electrical system and earth to locate the sections of the IT system with insulation faults. The insulation fault location system may consists of several functional modules combined in one device*

*Note: Insulation monitoring devices are defined in IEC 61557-8."*

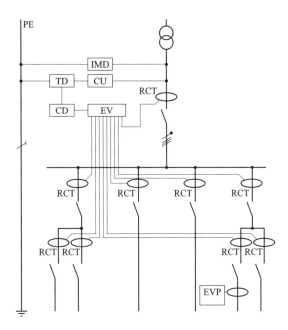

**Figure 10.1** Example of the components included in an insulation fault location system according to IEC 61557-9, Figure 1
*Legend*
*TD      Test Device*
*CU      Coupling Unit*
*CD      Control Device*
*EV      Evaluation Unit*
*RCT     Residual Current Transformer*
*IMD     Insulation Monitoring Device*
*EVP     Evaluation Unit, Portable*

# Literature

[10.1] American Society for Testing and Materials, ASTM International, 100 Barr Harbor Drive, P O Box C700, West Conshohokken, PA 19428-2959, USA

# 11 Technical Implementation of Insulation Monitoring Devices and Earth Fault Monitors

Nowadays, there is an insulation monitoring solution available for most unearthed systems. However, for technical reasons different measurement techniques are applied for different types of distribution systems. A distinction is made between

- Pure one or three-phase a.c. systems
- a.c. systems with directly connected rectifiers or thyristors
- d.c. systems

## 11.1 Insulation Monitoring of a.c. and Three-Phase IT Systems

For these systems a distinction is made between monitoring of the ohmic insulation fault and the measurement of the total system impedance to the protective conductor.

### 11.1.1 Measurement of Ohmic Insulation Faults

For a better understanding, the following additional explanations of the definitions regarding the core subject "insulation monitoring" are offered:

- $R_F$    insulation fault. An insulation fault is the value of one ohmic resistance to earth of the system being monitored.
- $\Sigma R_F$    absolute insulation resistance. The sum of all ohmic insulation faults to earth of the system being monitored.

The most common measuring principle is the superimposition of a d.c. measuring voltage between the system and the protective conductor. **Figure 11.1** shows this principle, which has been successfully used for decades. The measuring voltage is generated in the insulation monitoring device and applied to the system via high ohmic coupling resistances $R_i$. The simplified (schematic) drawing shows the monitored system as a conductor bar, because for the d.c. measuring voltage the secondary winding of the isolating transformer is low-resistance for the d.c. measuring voltage, it is superimposed on all conductors. On the occurrence of an insulation fault, the measuring circuit between system and earth closes over an insulation fault

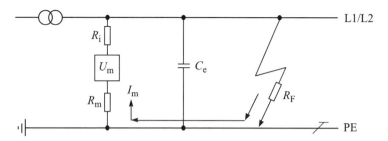

**Figure 11.1** Basic diagram of an insulation monitoring device

$R_F$, so that a d.c. measuring current $I_m$, proportional to the insulation fault, is in accordance with the following equation:

where

$$I_m = \frac{U_m}{R_i + R_m + R_F} \qquad (11.1)$$

$I_m$  d.c. measuring current
$U_m$  d.c. measuring voltage
$R_i$  internal resistance of the insulation monitoring device ($R_i$ results from the parallel connection of two times $R_I$ and other internal resistances)
$R_m$  measuring resistance of the insulation monitoring device
$R_F$  sum total of all ohmic insulation faults

The system leakage capacitances are charged to the d.c. measuring voltage and do not influence the measurement after the brief transient response. This technique measures the sum of all insulation faults. Devices based on this measurement method, which determine the absolute insulation resistance, are known under the protected trade name A-Isometer® (A for absolute, Iso for isolation), by the Bender company in Gruenberg, Germany. The a.c. components of the system are eliminated by filters.

A block diagram of an insulation monitoring device is shown in more detail in **Figure 11.2.**

The device is connected between system and earth via a measuring resistance $R_m$, the d.c. measuring voltage $U_m$, often a kΩ-meter and the coupling resistor $R_i$. If an insulation fault occurs, a voltage proportional to the insulation fault $R_F$ is measured across the resistor $R_m$. This voltage is compared to a pre-set or adjustable value of a trigger $T_1$.

If the preset response value is exceeded, an output relay K1 with voltage-free output contacts is activated via an amplifier stage V1. The circuits have a given hyste-

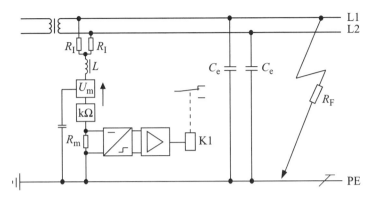

**Figure 11.2** Block diagram of an insulation monitoring device

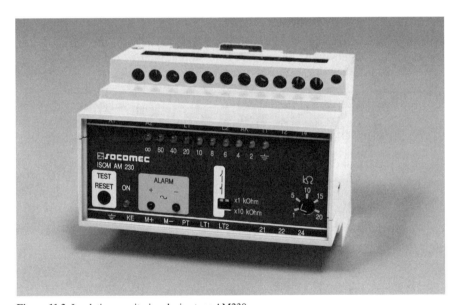

**Figure 11.3** Insulation monitoring device type AM230
[Photo supplied by the company Socomec, France]

resis in order to prevent the alarm relay from permanently triggering when reaching the response threshold. The kΩ-meter, which, in the circuit is positioned down-stream to $R_m$, indicates the absolute insulation resistance value of the system. A device according to this measuring principle is shown in **Figures 11.3**.

## 11.1.2 Measurement of the Leakage Impedance

Some countries measure the total leakage impedance of unearthed systems in medical locations as illustrated in **Figure 11.4**.

The principle of measurement is based on the superimposition of an a.c. measuring current between system and earth. The phase relation to the system changes permanently, for example by means of a fixed frequency close to the system frequency. At existing leakage impedances, this phase adjustment generates a superimposition on the "displacement voltage" between system and earth with a frequency corresponding to the differential frequency. By supplying a measuring current with constant amplitude, the amplitude of the superimposed voltage is directly proportional to the total system leakage capacitance to earth. A high ohmic coupling allows a sensitive evaluation and the injected measuring currents may be kept low. Thus, the fault current generated by the device for the medical location is less than 35 μA.

The analysis of the low-frequency superimposed voltage, in relation to the system, is done by filtering the voltage between the conductor and earth. In order to meet the requirements of NFPA 99 (USA), the maximum fault voltage of one conductor to earth present in a device is multiplied by the inverted value of the leakage impedance, so that the maximum prospective fault current in mA can be indicated.

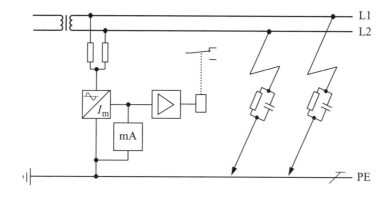

**Figure 11.4** Block diagram of impedance measurement

## 11.2 a.c. Systems with Directly Connected Rectifiers or Thyristors

The measurement techniques used for monitoring these systems are explained in the following chapter.

### 11.2.1 Measuring Method with an Inverter

A measuring principle similar to that described in Chapter 11.1.1 is used for monitoring a.c. systems with directly connected rectifier.

These systems (**Figure 11.5**) are often used for control systems of welding machines and presses as well as in the steel and chemical industry. These application fields often involve a.c. loads with directly installed rectifiers, such as magnetic valves or magnetic couplings. If insulation faults occur on the d.c. side of these loads, extraneous d.c. voltages will appear in the fault path in addition to the d.c. measuring voltage, as demonstrated in **Figure 11.6**. These d.c. voltages may be

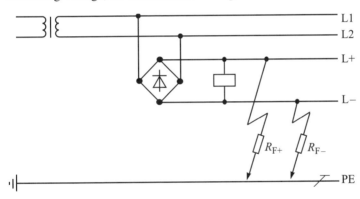

Figure 11.5 a.c. loads with rectifiers

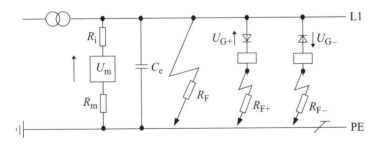

Figure 11.6 a.c. system with rectifiers, assumed insulation faults

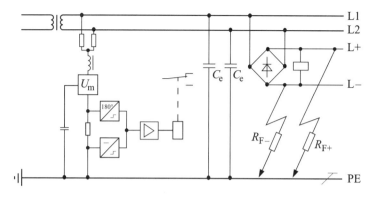

**Figure 11.7** Block diagram of an insulation monitoring device

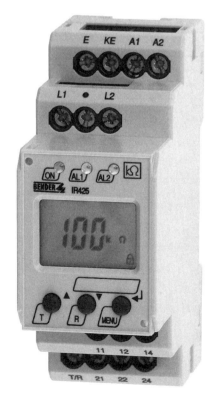

**Figure 11.8** A-Isometer® type IR 425 with adjustable response value for a. c. systems [Photo supplied by Bender, Gruenberg, Germany]

positive or negative depending on the location of fault and may falsify the insulation measurement.

This is the reason why, additional to the measurement method described in chapter 11.1.1, an inverter is integrated into the insulation monitoring device (**Figure 11.7**), in order to enable measurement of insulation faults on the positive as well as the negative side of the rectifier.

This measuring method requires a higher tripping sensitivity for faults on the d.c. side of rectifiers than for faults on the a.c. side. This ohmic insulation fault recognition has proved to be successful in practice as d.c. premagnetization of relays and contactors may jeopardize an a.c. control system. A $k\Omega$ indication is not possible with these devices. Such a device is illustrated in **Figure 11.8**.

### 11.2.2 Measurement by Pulse Superimposition

If for reasons of preventive maintenance it is sensible to indicate the insulation value in kW and to obtain an exact and quantitative measurement, independent of the type of fault and the location of the fault, it is recommended to use insulation monitoring devices with a measurement method other than the pulse superimposition [11.1]. This process involves the superimposition of a pulse voltage onto the system to be monitored (see Figure 11.12). This voltage is applied to high ohmic coupling elements between the system and the PE conductor. The pulse frequency is adjustable in such a way that it is aligned to the system leakage capacitances to earth (PE). The transient upper-range value of a measuring cycle is evaluated. The displacement voltage from the system is measured during the pulse pauses. In addition, in the subsequent phase of pulse superimposition the voltage fraction which is proportional to the fault is also measured. The obstructing displacement voltage is eliminated by means of the differential between two subsequent measurements. A direct voltage signal proportional to the fault is available for evaluation.

This measuring process enables the exact registration and indication of the insulation fault on the a.c. side as well as on the d.c. side. However, the system leakage capacitance is an important factor, since the pulse-duration of the measuring voltage has to be adjusted to the system leakage capacitances. This means that a longer response time must be accepted for these devices. In this measuring process symmetrical d.c. faults are also registered accurately.

## 11.3 d.c. Systems

Insulation monitoring devices or earth fault relays are also applied to d.c. systems.

### 11.3.1 Asymmetric Measurement

For voltages up to 230 V earth fault relays[1] are frequently applied. Most devices work with a bridge circuit (**Figure 11.9**).

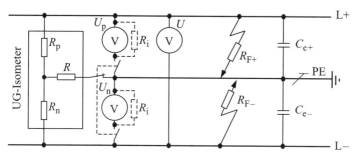

**Figure 11.9** Basic diagram of an UG-Isometer voltage asymmetry principle

Some versions are additionally equipped with a built-in or separate voltmeter to determine the insulation resistances $R_{F+}$ and $R_{F-}$. The measured voltage values may be applied in the following equations to calculate the insulation faults $R_{F+}$ and $R_{F-}$.

To be considered is that the earth fault monitor has to be isolated from the system during the voltage measurement and at the same time, the voltmeter has a certain internal resistance $R_i$. The voltages $U_n$ and $U_p$ are measured separately in succession:

$$R_{F+} = R_i \frac{U-(U_p+U_n)}{U_n} \tag{11.2}$$

$$R_{F-} = R_i \frac{U-(U_p+U_n)}{U_p} \tag{11.3}$$

The basic internal wiring of such a device is shown in **Figure 11.10**. The displacement voltage arising from an earth fault $R_{F+}$ (or $R_{F-}$) drives a measuring current $I_m$ which is recorded in an electronic measuring element. This activates the alarm relay

---

[1] An earth fault relay is a monitoring device that issues a signal on the occurrence of a single-pole insulation fault. Exact symmetrical insulation faults are not indicated.

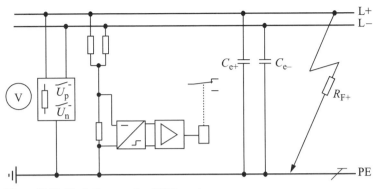

**Figure 11.10** Block diagram of an UG-Isometer

when the response value is reached. These devices do not need an auxiliary voltage as there is an electronic separation between the measuring circuit and the evaluating circuit. These devices are frequently applied in electronic control systems, because they do not superimpose an active measuring voltage on the system.

Indication in kΩ is not possible and exact symmetrical insulation faults cannot be detected with devices not equipped with an additional voltmeter. A device which has proved to be successful in practice is shown in **Figure 11.11**.

**Figure 11.11** UG-Isometer type UG207V [photo supplied by Bender, Gruenberg, Germany]

### 11.3.2 Measurement by Pulse Superimposition

Similar to a.c. systems with directly connected rectifiers or thyristors (see clause 11.2.2), the pulse method for measuring d.c. systems has the advantage, that the quantitatively exact kΩ-indication of the insulation condition of the system is possible. Furthermore, symmetrical faults can also be detected (**Figure 11.12**). In addition to the pulse measurement, another asymmetric process may also be superim-

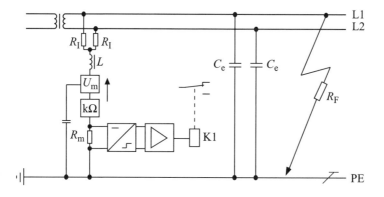

**Figure 11.12** Block diagram of the pulse measuring voltage

posed on these insulation monitoring devices for d. c. systems, so that selective fault location is possible. The devices have a LED to indicate "fault at +", "fault at –" and "symmetrical fault". Devices using the method of pulse measurement are also applied in d. c. systems.

## 11.4  Measuring Principles for the General Application in a. c. and d. c. IT Systems

More and more loads with galvanically directly connected rectifiers, thyristors or electronic converters are applied in both earthed and unearthed distribution systems.

Depending on the location of the insulation fault, for example the insulation fault comes after a downstream rectifier, the tripping conditions of the installed protective devices may be adversely influenced. In the earthed system, for example, residual-current circuit-breaker (RCCB) may be premagnetized by d.c. leakage currents, so that they do not trip or only trip in certain conditions.

Conventional insulation monitoring devices with d. c. voltage superimposition may only be applied on the precondition that the system conditions of the IT system are well known.

Both designer and user can profit from new measurement processes using microprocessors or micro-controllers, which allow a. c./d. c. as well as pure d. c. IT systems to be monitored reliably and easily.

Up to now there is no residual current protective device for TN and TT systems, which, on the occurrence of an insulation fault, can respond sufficiently exact, sensible and adequate as protection of persons in systems with d. c. and low-frequency

a. c. leakage currents. In this case the IT system can offer improved protection by insulation monitoring.

The following new measurement processes for insulation monitoring devices have been designed for universal application in a. c. and d. c. IT systems.

### 11.4.1 Microprocessor-Controlled AMP Measurement Process for the General Application in a. c. and d. c. IT Systems

Based on the pulse measurement method described in clause 11.2.2, a generally applicable insulation monitoring device has been developed using sophisticated and large-scale integrated micro-controller technology.

The "AMP" (**A**daptive **M**easuring **P**ulse) differentiates by using the switched-mode measuring voltage between system leakage currents, occurring as interference on the evaluation circuits, and the measured quantity proportional to the ohmic insulation resistance.

In applications with a. c./d. c. and d. c. IT systems generally broad-band disturbances are to be expected. Depending on fault conditions, converter generated leakage currents with frequencies between 0 Hz and the respective harmonic component of the system frequency. While it is easy to dampen interferences at system frequency or higher interferences by means of analogue filters, it is not easy to eliminate low-frequency interferences.

Adapting filters have been integrated into the software of new AMP measurement technique. In this way low-frequency interference, such as those from frequency converters, can be controlled. **Figure 11.13** shows a typical device, wich is based on this principle.

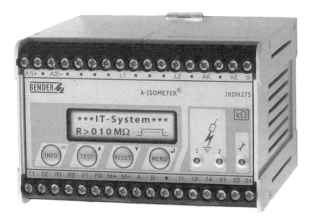

**Figure 11.13** Insulation monitoring device, A-Isometer® IRDH275 [Photo supplied by Bender, Gruenberg, Germany]

New evaluation software allows flexible and automatic adaptation of the measuring voltage and measuring time to existing system conditions. The adjustment parameter such as response value(s), special alarm and display functions, are programmable and are stored in a non-volatile memory. These devices are equipped with a RS 485 interface to fulfil the future requirements of power system management.

The measurement methods of the universally designed devices reach their limits in applications where converters are susceptible to high interference stress, for example in heavy machinery, such as brown coal excavators. If higher system voltages and extremely low-frequency control procedures are involved, a special frequency-code measurement method has been developed, which is explained in the following.

### 11.4.2 Microprocessor-Controlled Frequency-Code Measurement Method for IT Systems with Extreme Interference

In IT systems with high-performance converter drives, voltage components in the frequency range of 0 Hz to 10 Hz with amplitudes of several hundred volts may occur.

These voltages are superimposed on the measuring voltage of an active insulation monitoring device as an interference voltage.

If these interference voltages exceed the maximum permissible value of the device, measurement suppression or false measurements may be possible reactions of a conventional monitoring device. In particular, with difficult applications, for example with the insulation monitoring of the control systems of a bucket wheel excavator in brown coal open-cast mining, a measuring method has been developed that provides improved and more reliable operation even under difficult conditions.

There are two underlying basic ideas to the frequency-code method:
- narrow-band measuring and
- measuring frequency adaptation

An insulation monitoring device operating according to the frequency-code method based on micro-processor techniques has been developed. This means that most functional parts of the device are software modules.

After setting the duration of the measuring voltage against earth for the IT system, a specially designed software module based on the Fourier analysis determines the shape of the measuring voltage so that the desired spectral parts receive an amplitude as high as possible. The mid-frequencies of digital filters are adapted to these spectral parts. The supplied measuring voltage causes a current flow via the system leakage capacitance, insulation resistance and coupling circuit. The resulting voltage drop at one part of the coupling resistance is fed into an analog-digital converter (ADC). The digital band-passes are being fed with the output data of the ADC.

The r.m.s. value ($U_{rms}$) for one period of the measuring signal is determined from the output data of the filter. The system leakage capacitance and the system insulation resistance can be calculated by the quotient of the r.m.s values and the quotient of the filter mid-frequencies. An insulation device using the method of frequency-code measurement, for example, is installed into the systems of motor drives of the bucket wheel excavator 292 (**Figure 11.14**) from Rheinbraun.

**Figure 11.14** Bucket wheel excavator 292 [Photo supplied by Rheinbraun company, Germany]

## 11.5 Insulation-Fault-Location System in a.c. and d.c. IT Systems

A special advantage of the IT system in regards to improved operating safety is the possibility to detect and to eliminate insulation faults during operation.

In contrast to earthed TN and TT systems, an insulation fault or even a direct single-pole earth fault will not trip the fuse. Selective disconnection of a part of the system by an overcurrent protective device only happens in the event of a double fault to earth, which is rarely the case (simultaneous low-ohmic earth fault of at least two phase conductors).

Insulation monitoring devices continuously measure the insulation resistance of the entire galvanically connected IT system, including all connected consumers. In general practice this means, the parallel connections of the insulation resistances of each phase conductor to earth are measured. The actual total insulation resistance can be indicated on a k$\Omega$-meter. If the value falls below a response value which can usually be set and adapted to the existing system conditions, the insulation monitoring device signals insulation faults.

Modern insulation-fault location systems not only allow the detection of direct earth faults, but also the selective detection of comparatively high ohmic insulation faults.

Comparing the costs of such insulation-fault location systems with the costs incurred using conventional fault location methods, by disconnection of sections, plus adding the costs from interruptions to operations, for examples the production process, it becomes obvious that installed fault location systems with their added safety aspects, are of great economical benefit.

The advantage of operating safety, which the IT system has over earthed systems, can only optimally be used in combination with insulation-fault location systems. Of course, the existing system conditions, number and type of consumers etc. need to be considered as well.

IT systems can continue to operate even if they indicate an insulation fault. However, regulations for erecting such systems require that the reported fault be eliminated as soon as possible.[2]

Various methods are used for the location of insulation faults. All common methods are based on the residual current measuring principle, where a test current from an external source or from the system voltage flows via the fault location against earth. It can be evaluated with adjusted residual current transformers or current clamps.

In the following, three methods of insulation-fault-location for IT systems are described.

---

[2] See also IEC 60364-4-41:2005-12, 411.6.3.1 and Chapter 4

### 11.5.1 Insulation-Fault-Location Systems for d. c. IT Systems

The evaluating units of these locations systems are based on the principle of residual current detection. Residual current detection in d. c. systems is possible with the compensation measuring method. The d. c. magnetization of a residual current transformer is compensated by means of an injected a. c. current via an additional winding.

As is generally known, in a fault-free system the total of the operating currents in the supply line always equal zero (Kirchhoff's law). Insulation faults in unearthed a. c. IT systems cause residual currents as an effect of the existing systems capacitance.

In d. c. IT systems an artificial test path is created to return the d. c. leakage current. This task is carried out by the test device G1 (**Figure 11.15**). the operating current is routed by a residual current transformer. Insulation faults occurring downstream of the assigned current transformer generate a residual current, which is evaluated in the built-in electronic circuitry and a signal "insulation fault downstream of the current transformer" is indicated.

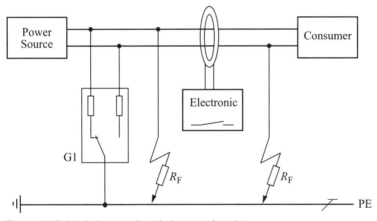

**Figure 11.15** Block diagram of residual current detection

### 11.5.2 Insulation-Fault Location Systems for a. c. and d. c. IT Systems

Active and passive measuring methods are used for insulation-fault location in a. c. and d. c. IT systems. Active measuring procedures superimpose a test current with a frequency of usually only a few hertz between system and earth and in that way fault currents can be easily detected and evaluated by filters. In most cases, however, these methods cannot be used in systems involving high leakage capacitances since the current to be evaluated is not only determined by the ohmic insulation fault but also by the leakage capacitance.

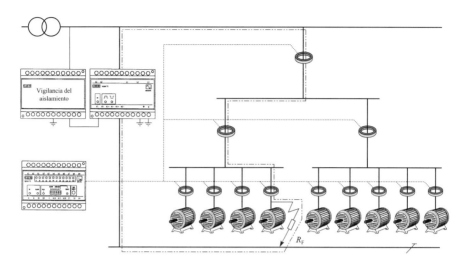

**Figure 11.16** Example of an insulation-fault location system for automatic detection of faulty system parts

Passive methods generate a test current driven by the system being monitored, which can be evaluated selectively. In automatically operated systems, fault location is activated by the alarm contact of an insulation monitoring device. A test device starts generating a test current as soon as the insulation monitoring device signals that the value is below the preset response value.

Similar to the method described for d. c. IT systems in 11.5.1, a test device generates a test current alternately between each phase conductor and earth, which flows back to the system via the insulation fault and earth where it can be detected selectively by means of special residual current transformers with evaluators (**Figure 11.16**)

### 11.5.3 Portable Insulation-Fault-Location System for a. c., d. c. and Three-Phase IT Systems

There are situations where the general operating conditions rule out the use of stationary insulation fault location systems. In these cases, the use of portable insulation fault location systems is recommended.

The insulation fault location system described here is being decoupled between the system to be monitored and the PE conductor and adapted to the system. In the event of an insulation fault, a leakage current will run from the location of the fault to the intact conductor, and generates a d. c. current in the test device. This pulsating d. c. current is magnetically evaluated by means of a measuring clamp and converted into a visible and audible signal by the electronic unit connected to the measuring clamp. A typical device is illustrated in **Figure 11.17**.

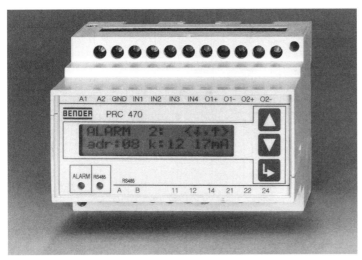

**Figure 11.17** Control device for insulation-fault-location system, type PRC470 (Photo supplied by Bender)

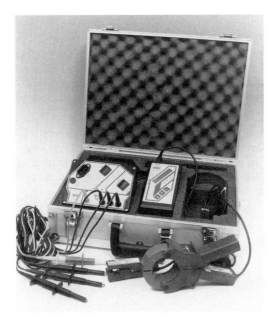

**Figure 11.18** Portable insulation-fault location system EDS3065 [Photo supplied by Bender, Gruenberg, Germany]

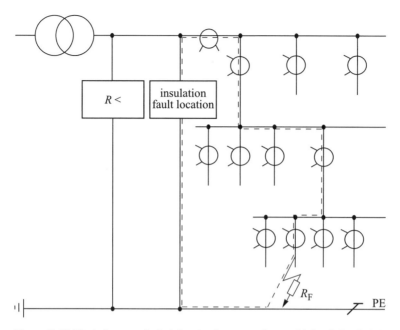

**Figure 11.19** Block diagram of a fault location by means of a portable insulation-fault location system

For fault location, the test device is connected at an optional place between the system and the protective conductor, but the best place is the main distribution point. Each outgoing cable is then grasped by the current clamp. A visible and audible signal at the evaluating unit indicates that the insulation fault is further downstream of the measuring point. This fault shall then be pursued. **Figure 11.19** illustrates the principle of the fault location procedure.

By means of a selective insulation fault location system, i. e. exact fault location of high-ohmic insulation faults in an unearthed system, another step has been taken towards improved operating safety in electrical installations. The combination of insulation monitoring devices and selective insulation fault location systems can increase the operating duration of an unearthed power distribution system.

## 11.6 Summery

This chapter attempted to provide an indication on a variety of measuring methods involving insulation monitoring devices. Insulation monitoring devices are available on the market for nearly all unearthed IT systems. They range from simple devices for control systems to complex device for converter drives. For IT systems up to 1000 V a.c. and 1500 V d.c. these devices must comply with IEC 61557-8. There are also suitable devices available for insulation-fault location systems, which should comply with IEC 61557-9.

Up-to-date insulation monitoring devices already have integrated control and test devices for insulation fault detection. Via interface and available data transmission, systems are able to remotely detect fault locations.

For special applications a number of particular devices are being offered. There is still research done in this field. The future might hold completely different measuring methods.

## Literature

[11.1] Junga, U.; Kreutz, W.: Erdschluss- und Isolationsüberwachung in Gleichspannungsnetzen. etz-b Elektrotech. Z., Ausgabe b, Bd. 29 (1977) H. 4, S. 125 – 128

# 12 Response Values of Insulation Monitoring Devices (IMDs)

The selection of the appropriate response value for the insulation monitoring device is often complicated, since they can be applied in diverse environmental conditions, e.g. in an hygienic distribution board of a hospital with high requirements on the insulation values or in systems for smelting furnaces with low insulation requirements.

The insulation condition of an electrical installation naturally depends on many different factors. The mode of operation, ambient conditions and lifespan of an installation are all important factors.

A number of standards and regulations with recommendations for the respective response values for ohmic insulation values are listed in **Table 12.1**. Values given are minimum values, but insulation monitoring devices should be set with a value approximately **50 % above the recommendation** in order to make allowance for the permissible tolerances of the devices. If the indication value for example requires 50 k$\Omega$, the setting on the device should therefore be 75 k$\Omega$. Insulation monitoring devices selected for the measurement of the ohmic insulation resistance should be in compliance with IEC 61557-8 (**Figure 12.1**).

**Figure 12.1** Insulation monitoring device type IRDH1065 [Photo supplied by Bender, Gruenberg, Germany]

| Standard | Application | Nominal Voltage | Required Insulation Resistance | Recommended response value of the IMD |
|---|---|---|---|---|
| IEC 60364-5-53 Annex H | Selection and erection of electrical equipment – Isolation, switching and control | 230 V | pre-warning 100 Ω/V= 23 kΩ; warning 50 Ω/V= 11.5 kΩ | 34.5 kΩ<br><br>17.25 kΩ |
| IEC 60364-7-710 | Medical locations | 230 V | 50 kΩ | 75 kΩ |
| UL 2231-2 | Supply Circuits for electric vehicles | | 100 Ω/V at nominal voltages | |

**Table 12.1** Required insulation values and recommended response values of insulation monitoring devices (IMDs)

**Table 12.2** shows the response time for insulation monitoring devices for the monitoring of pure a.c. systems or for galvanically connected d.c. circuits and d.c. systems according to IEC 61557-8.

According to IEC 61557-8:2006 the following definitions apply:

| System | Response Time $t_{an}$ | Maximum Response Value $R_{an}$ | System Leakage Capacitance $C_e$ |
|---|---|---|---|
| a.c. | ≤ 10 s | 0.5 | 1 μF |
| d.c. | ≤ 100 s | 0.5 | 1 μF |

**Table 12.2** Response times for insulation monitoring devices according to IEC 61557-8

**Response time ($t_{an}$)**: Time required by an insulation monitoring device to respond under the conditions specified in 6.1.2.
(With a leakage capacitance $C_e$ of 1 μF and at the nominal system voltage, the insulation resistance shall be suddenly reduced from nearly infinity to 50 % of the minimum response value $R_{an}$ and the delay to the operation of the output circuit shall be measured; IEC 61557-8, 6.1.2.)

**System leakage capacitance ($C_e$)**: Maximum permissible value of the total capacitance to earth of the system to be monitored, including any connected appliance, up to which value the insulation monitoring device can work as specified.

# 13 Physics of the IT System

## 13.1 Leakage Currents in the IT System

An essential factor for the right selection and setting of insulation monitoring devices is the value of the total leakage capacitances of the monitored system [13.1]. As these values of the electrical installation are generally unknown, the determination of the leakage capacitance will be explained in this chapter. The leakage current of the a. c. IT system is directly proportional to the system leakage capacitance.

The total system leakage capacitance $C_{e\,tot}$ is composed of the capacitance of the transformer T between secondary and primary winding, the conductive capacitance of the component conductors $C_e$ against PE as well as the capacitance $C_L$ of the system conductors to earth within the current-using equipment.

In most cases the transformer capacitance may be negligent. For modern transformers it is approximately between 5 nF and 30 nF, depending on the power rating.

Phase-to-phase capacitances are dependent on the strength of the insulation between the conductors, distance h, the matter constant $\varepsilon_r$, the electric constant $\varepsilon_0$ and the area of the insulation between the conductors. Generally in standard installations only the capacitance of the conductors to earth determine the capacitances of the system. This is further explained in **Figure 13.1**.

The capacitance is calculated according to the following equation:

$$C = \frac{2\pi \varepsilon_0 \varepsilon_r l}{\ln(2h/r)} \qquad (13.1)$$

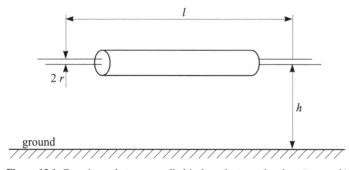

**Figure 13.1** Capacitance between a cylindrical conductor and a plane (e.g. earth)

The following equation is an example for calculating the capacitance of a cable:

Distance $\quad h = 50$ mm
Length $\quad l = 500$ m
Radius $\quad r = 0.69$ mm
Matter constant $\quad \varepsilon_r = 1$ (air)
Electric constant $\varepsilon_0 = 0.8855 \cdot 10^{-13}$ F/cm

Capacitance $\quad C = \dfrac{2\pi \cdot 0.885 \cdot 10^{-13} \text{F/cm} \cdot 1 \cdot 500 \cdot 10^2 \text{cm}}{\ln(2 \cdot 50\,\text{mm}/0.69\,\text{mm})}$

$C \approx 5.5$ nF

Inductive coupling in most cases can be neglected.

Inductive components of the single conductors in wires are compensated in symmetrical order within the static mean, so that only voltages in the mV-range are induced for example between start and end of a PE conductor.

### 13.1.1 Calculation of Leakage Currents in IT Systems

For the calculation of leakage currents in the IT system, the sum of the leakage capacitances must be considered. They can be determined by calculation or measurement. If the required factors are known, the equation (13.1) can be applied.

Measurement of the leakage capacitances can be done according to various methods (integrated insulation monitoring devices should be de-energized during measurement).

### 13.1.2 Determination of the Leakage Capacitances in the De-Energized System

Leakage capacitance $C_e$ can be determined by means of an a. c. power source $U$ and by measuring $U_C$ and $I_C$ (**Figure 13.2**) according to the following equation:

$$C_{e\,tot} = \dfrac{|I_C|}{|U_C|\omega} \tag{13.2}$$

In the single-phase a. c. system the total capacitances $C_{e\,tot}$ equals 2 $C_e$, in the three-phase 3 $C_e$, in the four-phase system 4 $C_e$.

The fact should be considered that $X_{Ce} \ll R_i$ of the voltmeter, so that the measuring uncertainty remains low, for example when $R_i$ equals 10 MΩ, $C_{e\,tot}$ should be > 0.3 nF.

Furthermore it is very important that the insulation resistance of the systems to PE is higher than the capacitive impedance (determination, for example by indication of the insulation monitoring device in operation).

If the insulation resistance has a finite value, the resistance fraction $R_F$ can be determined with a d. c. voltage source by means of the described measurement.

Measurement with the a. c. voltage source indicates the leakage impedance $Z_e$.

From both measured values the leakage capacitance can now be determined by the following equation:

$$C_{e\,tot} = \frac{1}{\omega\sqrt{Z_e^2 - R_F}} \qquad (13.3)$$

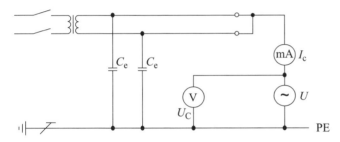

**Figure 13.2** Measuring circuit to determine the capacitance in a disconnected circuit
U     a. c. measuring source
mA   Milliampere meter (mA-meter)
V     High ohmic voltmeter ($R_i > 10$ MΩ)

### 13.1.3 Determination of the Leakage Capacitances in the Energized System

It is recommended that the insulation monitoring device should likewise be deenergized during measurement (**Figure 13.3**).

The leakage capacitance of the system can be calculated by means of the measured current. The mA-meter will cause a direct low ohmic insulation fault in the system. However, before this is done, it should be ensured by means of the insulation monitoring device that the insulation value of the system is sufficient. In addition the voltage of all phase-conductors to earth should be checked. If these voltages are symmetrical and about the same value as half of the system voltage, it is confirmed, the system is a fault-free IT system. But even in the fault-free IT system, slight asymmetric voltages of the conductors to earth may occur, caused by asymmetrical leakage capacitances.

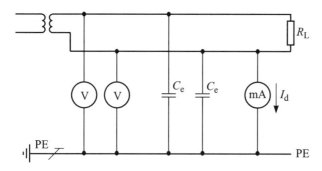

**Figure 13.3** Measuring circuit to determine the capacitance during operation

The leakage current measurement can now be carried out. The occurring voltage for a. c. and three-phase systems can be determined according to the following equations:

a. c. systems:
$$I_d = U\omega C_e \tag{13.4}$$

leakage capacitance:
$$C_e = \frac{I_d}{U\omega} \tag{13.5}$$

For three-phase system:
$$I_d = U\sqrt{3}\,\omega C_e \tag{13.6}$$

leakage capacitance:
$$C_e = \frac{I_d}{U\sqrt{3}\,\omega} \tag{13.7}$$

where:

$C_e$      leakage capacitance of each conductor to earth
$U$      phase-to-phase voltage

In this case, the respective leakage capacitance of a conductor to earth should be applied for $C_e$; assuming that all leakage capacitances for each conductor are equal.

## 13.2 Voltage Ratio in the a. c. IT Systems

In IT systems the voltages of the phase conductors against earth are proportional to the leakage impedances of the system. These impedances consist of the capacitances of the conductors and of the electrical equipment to earth and the hereby parallel connected insulation resistances. If these leakage impedances are equally high for each conductor, all phase conductors also have the same voltage against earth. Voltmeters with high internal resistance which are connected between the phase conductors and earth show the same value. In three-phase systems this is the phase-to-neutral voltage, in a. c. systems half of the phase-to-phase voltage is indicated. Insulation monitoring devices should therefore be coupled symmetrically.

If a low ohmic insulation fault occurs is in one of the conductors, the voltage against earth will decrease. As the voltage between the conductors remains, the intact conductors are lifted to the level of the conductor voltage to earth – thus, the conductor voltage [13.2].

It must be considered that when there is a fault to earth in one conductor in an unearthed system, the neutral point of the transformer takes over the phase voltage and the intact phase conductors are raised to the phase-to-phase voltage against earth – thus, the phase conductor voltage.

This increased demand on the voltage at the point of reduced electrical resistance may cause a break-down, which may lead to a double short-circuit with a high current resembling a short-circuit. As a result the up-stream protective device is triggered and hence the supply is interrupted. For this reason it is advisable to detect a first insulation fault when developing, locate the fault and to switch off, before an **unexpected** failure of the supply. This necessitates to continuously monitor the insulation resistance of the system with an insulation monitoring device. Exceeding the minimum value of the insulation resistance is to be signalled visually or acoustically; in special circumstances switching-off the installation may even be required.

The earth-fault current $I_d$ occurring at the fault location is determined by the leakage resistance of the intact conductors and their voltage in the event of a fault. Contrary to protective separation, earth-fault currents in extended systems may become so high that they can cause a dangerous current, when the phase-conductor is directly touched.

Quite often it is erroneously assumed, that it is possible to directly touch a phase-conductor in the IT system. In a four-wire system the N-conductor is raised to the level of the neutral voltage against earth in the event of a low ohmic insulation fault in a phase conductor. The load connected to such a system must be able– at least temporarily – to withstand the conductor voltage against earth without damage (**Figure 13.4**).

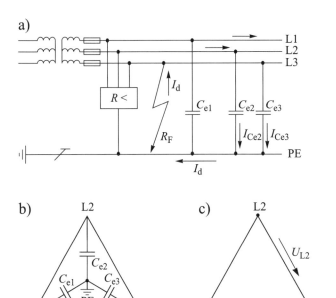

**Figure 13.4** Voltage and current ratio in the IT system
a) IT system with a short circuit to earth in conductor L3. The earth fault current $I_d$ flows via the capacitances of the intact conductors.
b) Conductor voltage in the event of symmetrical phase-to-earth capacitance. All conductors carry the phase-to-neutral voltage against earth.
c) Conductor voltage against earth in the system. IT system with a short circuit to earth in the conductor L3. The intact conductors carry the phase-to-phase voltage against earth which determines the magnitude of the earth fault current via the phase-to-earth capacitances.

In contrast to the a. c. IT system, the voltage to the protective conductor in the d. c. IT system is not determined by the value of the leakage capacitances. The voltage between system and earth equals the ratio between plus and minus against earth of the insulation resistance. In well maintained systems this voltage will be half the voltage of the phase-conductor.

## 13.3 Overvoltage in a. c. and Three-Phase a. c. IT Systems

International and national standards, as well as publications [13.3, 13.4, 13.5, 13.6] only give general advice and recommendations on the concept of IT systems. The International standard IEC 60364-4-41, for example, does not allude to earthing by high impedance, in order to lower overvoltage; neither are statements by experts a common thing. In order to achieve effective insulation monitoring – in practice – the live parts against earth preferably are being isolated. Results of theoretical and mathematical investigations [13.7] of transient overvoltage in a. c. and three-phase a. c. IT systems up to 1 000 V are supporting planning and operations of electrical installation [13.8].

| Causal System Process | IT System $ü$ | TN/TT System $ü$ |
|---|---|---|
| Low ohmic insulation fault, normal (balancing processes) | 2.5 | 1.0 |
| Low ohmic insulation fault, intermittent | 4.5 | 1.0 |
| Low ohmic insulation fault in the supplying system | 4.0 | – |
| Switching from no-load inductivities, especially transformers | 5.5 | 3.0 |
| Switch-off of no-load wires (fault-free) | 2.0 | 2.6 |
| Switch-off of wires (low ohmic insulation fault) | 3.0 | 4.3 |
| Overvoltage by resonance and harmonics | 3.0 | – |
| Switch-off of low ohmic insulation faults | 3.4 | 3.0 |

**Table 13.1** Maximum overvoltage factor y for different systems

Overvoltage in a. c. and three-phase a. c. systems not only just describe a multi-level problem, but are an important topic in terms of safe and trouble-free systems operation as well as secure operation of equipment. Especially in IT systems (systems with insulated neutral point), which are engaged in the supply of sensitive applications in industry and the health sector, for example in operating theatres, the sources of voltage rise, the degree of overvoltage (overvoltage factor y, **Table 13.1**) and damping are an interesting subject.

Most cases of voltage rise can be traced back to a switching process (operational or remote) on electrical resonant circuits. Certain negligence, simplification and condensation may be useful and are permitted. It is sufficient to base a simple resonant circuit with $L$, $C$, $R$ and $t$.

### 13.3.1 Sources of Overvoltage

The single-pole insulation fault is characterized by the following layout: a live conductor is directly connected to earth or PE conductor, via a resistance or via an elec-

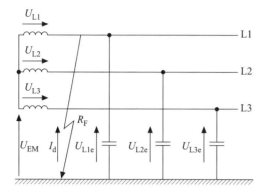

**Figure 13.5** Equivalent circuit diagram of a low ohmic insulation fault

tric arc (**Figure 13.5**). In IT systems restricted in space the low ohmic insulation fault arcs, which may occur, switch-off by themselves. There are no damages done at the fault location. The system may remain operational over a prolonged period of time with an low ohmic insulation fault, without influencing the energy transmission, if the insulation of the fault-free conductors are intact and can resist both the transient overvoltage and the stationary voltage rise.

A low insulation level may result in two low ohmic insulation faults, which have to be switched off immediately. The stationary voltage rise – at three-phase with neutral a. c. IT systems, it is to the level of the interlinked voltage – is no problem for the normal high insulation level, because the requirements for the determination of the creepage distances and clearances demand high safety levels (factor 3 to 5).

The overvoltage arising from stressed insulation often accelerate the aging process and leads to a break-down of the already weakened insulation.

Sub-harmonics occurring after the elimination of the low ohmic insulation fault possibly result in intermittent low ohmic insulation faults. The voltage recovery, occurring approximately 5 ms after eliminating the low ohmic insulation fault, may also contribute to re-ignition and, in special cases, to intermittent low ohmic insulation faults.

### 13.3.2 Transient Phenomena at Single-Pole Low Ohmic Insulation Faults

Transient phenomena may be classified into three processes, which occur in sequence and in circuits, to be examined separately. The discharge process of the phase-to-earth capacitances of the low ohmic insulation faulted conductor applies immediately in accordance with the predetermined travelling-waves. This high-frequent discharge process is irrelevant for the overvoltage, because the voltage is going to equal zero.

The charging process of the distributed capacitances of the fault-free conductors applies at the same time as the discharge process, but with a lower frequency. Maximum overvoltage occurs here, if the low ohmic insulation fault happens at zero crossing of the stationary earth-fault current. The value of the overvoltage, which is theoretically probable at first resonance in the three-phase a. c. system is 2.5 times, respectively in the a. c. system 1.5 times the peak value of the voltage.

The voltage recovery at the fault location, as well as through the low ohmic resistance related resonance process of the phase-to-earth voltage of the intact conductors, have peak value, if the light arc of the low ohmic insulation fault has gone out at zero crossing; because of the phase shifting the neutral point displacement voltage has reached its peak and remains as direct voltage on the capacitances of the intact conductors. The voltage rise may be double the peak value of the phase-to-earth voltage at normal operation, but do not harm the intact insulation, because type tests for example according to EMC-requirements need a much higher test voltage. The impedance of the insulation monitoring device, a probable impedance earthing and the insulation level of the system call for a reduction of the overvoltage.

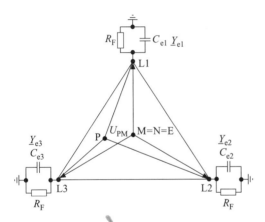

**Figure 13.6** Equilateral voltage triangle of the phase-to-phase and phase-to-earth voltage in the practical system
Where

| | |
|---|---|
| $R_F$ | Insulation resistance |
| $C_{e1-e3}$ | Conductor – earth – element capacitance |
| $Y_{e1}$ | Conductor – earth – admittance |
| $U_{PM}$ | Shift voltage |
| M | Neutral point |
| N | Neutral point of the system |
| E | Potential point (reference point zero) |
| P | Centre of the system by asymmetrical conductor – earth – admittance (special case: low ohmic insulation fault / single-pole insulation fault) |

### 13.3.3 Stationary Voltage Rise

In the healthy system an equilateral voltage triangle is aroused from symmetrical, stationary voltage against earth and their derivations (insulation resistances and capacitances) (**Figure 13.6**). Its star point is the neutral point, the electrical centre of gravity of the system and, theoretically, also the earth potential. The fault-free system only has very low voltage system asymmetries, on the contrary however systems with single- or double-faults have significant system asymmetries.

In the single-pole stationary fault of the three-phase a. c. system the phase-to-earth-voltage of the active conductor, afflicted with the low ohmic resistance fault, breaks down in dependence of the fault resistance and becomes zero at a "dead earth-fault". The phase-to-earth voltages of the earth-free live conductors, on the contrary, rise to the maximum of the diametric voltage; and the neutral point displacement voltage rises at the most to the phase-to-earth voltage. The actual voltages for the particular low ohmic insulation fault case may be calculated by means of the accepted curves.

The same considerations apply for the single-phase a. c. system, however a doubling of the phase-to-earth voltage in relation to the total system voltage has to be considered at the low ohmic insulation fault conditions.

In a. c. and three-phase a. c. systems the maximum phase-to-earth voltage remains below the rated voltage of the equipment and may be negligent at the selection of the equipment. The contrary with three-phase a. c. IT systems, which should not really be chosen. If the neutral conductor is part of the IT system, the operational equipment, which is connected between the outer conductor and the neutral conductor, must be insulated and suitable for phase-to-phase voltage.

Often the problem with three-phase with neutral a. c. systems is thoughtlessly regarded by standards and expert literature, for example in IEC 60950, as a general problem with IT systems.

### 13.3.4 Intermittent Low Ohmic Insulation Fault

In IT systems, which are not largely spread (fault current $I_d$ <1500 mA) is hardly any danger of intermittent low ohmic insulation faults, because the possible light arc at the zero crossing of the current extinguishes itself. However, for IT systems with insulated neutral point, overvoltages have been calculated up to twice the peak value of the phase-to-earth voltage. Theoretically this does not exclude a repeated low ohmic insulation fault. The following distinctions have to be made:

- The low ohmic insulation fault remains (see clause 13.3.3)
- The low ohmic insulation fault is extinguished at the zero-crossing of the capacitive, stationary earth-fault current. At worst case (free neutral point) one has to expect overvoltage factors of $y < 4$ are to be expected

- The low ohmic insulation fault is extinguished permanently and re-ignites according to second and third conductive coating. The maximum calculated overvoltage factor is $y < 7.5$

The $y$-factors are theoretically calculated maximum values, which are never achieved in practical experience, because the system load, the finite values of the insulation level and the impedance of the insulation monitoring device, which is always there, have dampening effect.

### 13.3.5 Insulation Faults in Supply Systems

Decisive factor for the voltage rise in IT systems, caused by insulation faults in the supplying system, is the nominal system voltage and the handling of the neutral point of the upstream system. The most unfavourable constellation – supplying system with insulated neutral point or inductively earthed – the following stationary voltage rise applies:
- supply from a low-voltage system: maximum of 1.5 times the neutral point of the phase-to-earth voltage of the IT system
- supply from the medium voltage (MV): maximum of 4 times of the neutral point of the phase-to-earth voltage of the IT system

It can be assumed that the voltage transmission is purely capacitive and can be explained with the so called "equivalent zero circuit diagram" of the transformer. The impedances of the wires have a high dampening effect on the value of the overvoltage of IT systems.

### 13.3.6 Switching of Inductivities

Off-load transformer, if separated from the system on the part of low-voltage, respectively connected to the no-load system, are an almost pure inductive load. The switching process may cause an overvoltage, due to the phase-angle of the magnetising current of the weak light arc and with the possible chopping of the no-load current. The inductivities and capacitances of the transformer create resonant circuits, which are energized by the switching process.

Basis for the overvoltage-calculations is "energy", meaning the consideration of energy of the system, which is to be switched-off. Of importance is the number of phases of the transformer, the handling of the neutral point on the primary side of three-phase a. c. transformer and the point of the break-time (zero-crossing or maximum) of the no-load current. During the switching process at zero-crossing the overvoltage factor was calculated with $y < 2$, the secondary system did not show any dampening effect, leading to higher overvoltages during switching in the maximum current. These overvoltages however, were dampened strongly by the connected system.

### 13.3.7 Switching of Wires and Capacitors

The switching of no-load wires and capacitor batteries (for reactive-current compensation) is another source for overvoltages, because they are engaged as energy storage in the harmonic processes of the circuit.

When the wires and capacitors are turned-on, overvoltage factors $y < 3$ are created in a. c. and three-phase a. c. systems. The capacitance ratio "total system to sub-systems" and no longer existent residual loads on the lines, lower the factor to $y < 2$.

When the wires and capacitors are turned-off, the overvoltage factors are tied to different marginal conditions. Shutdown cycles without restrikes and interruptions of the capacitive current at or before zero-crossing are regarded as being free of overvoltages, at negligible system expansion, sub-system disconnection and observations of the capacitance ratios.

### 13.3.8 Resonance and Harmonics

In general resonance is not a transient process. They may occur as a result of switching processes, but they do not die down in the sense of a transient overvoltage. They remain until the conditions are changed. Resonance in an electrical system always depend on the capacitive and inductive system components and occur generally in medium and high voltage systems.

In case of harmonic resonance in a three-phase a. c. system with asymmetrical operating conditions, for example single or two-phase open circuits, the overvoltage factor is $y < 3$. Here a maximum length of the wires may be determined, in order to exclude overvoltages.

Harmonic currents create harmonic voltages on the impedances of the supply system. If they are needed it is advisable to examine the impedances in regards to the respective harmonic frequencies. a. c. and three-a. c. systems with negligibly expansion will not show any places with resonance.

### 13.3.9 Voltage Rise at Short-Circuit Disconnection

If a short-circuit is disconnected by switches or fuses, transient recovery voltages above the poles may occur, due to the energy exchange.

In a. c. systems (rated voltage $\leq$ 230 V) there are no or very low overvoltages, because of the installation of small wires. In three-phase-a. c. systems (rated voltage $\leq$ 500 V) overvoltages which fall below a value of twice the phase-to-earth voltage, are to be expected. In this case, however, the location of the short-circuit has to be considered (bus bar short-circuit or short-line fault).

The eventuality of dampening of overvoltages in unearthed IT systems by high resistances between system and earth may be applied in those cases, where the advantages of continuous monitoring may be forgone.

## 13.4 IT Systems and the Second Fault

The only standard that mentions the second fault in IT systems is IEC 60364-4-41: 2005-12, 411.6.1. This is surprising, since the occurrence of second faults is usually not expected in the requirements of standards. The reason of this is found in IEC Guide 104 of IEC (3$^{rd}$ edition, 1997-08) [13.9] which says:

"The simultaneous occurrence of two independent and unrelated faults need not normally be taken into account, because of the likelihood of such an event is so low that the risk is generally at a tolerable level.

*Note: The single fault criterion is used extensively in relation to hardware failures in the field of electrical safety to provide protection against electric shock. ... [5.3.5]"*

The following is an attempt to explain the specific characteristics of IT systems in regards to the second fault as well as making clear the difference between insulation faults in general and the first and second fault in particular.

IEC 60364-4-41:2005-12, 411.6.1 is mentioning the protection against electric shock under fault conditions:

"Provisions shall be taken, however, to avoid risk of harmful pathophysiological effects on a person in contact with simultaneously accessible exposed-conductive-parts in the event of two faults existing simultaneously." And:

"After the event of a first insulation fault, conditions for automatic disconnection of supply in the event of a second fault occurring on a different live conductor shall be as follows:..." [411.6.4]

"In cases where an IT system is used for reasons of continuity of supply, an insulation monitoring device shall be provided to indicate the occurrence of a first fault from a live part to exposed-conductive-parts or to earth. This device shall initiate an audible and/or visual signal which shall continue as long as the fault persists. If there are both audible and visible signals, it is permissible for the audible signal to be cancelled.

*Note 1: It is recommended that a first fault be eliminated with the shortest practicable delay." (411.6.3.1)*

Hence the second insulation fault in an IT system with negligible impedance and a low-resistance, conducting connection of an outer conductor or the neutral conductor with earth or earthed parts.

Furthermore for the determination of the maximum disconnecting time according to Table 41.1[1] the assumption has been made, that the first insulation fault and the

---

[1] See IEC 60364-4-41:2005, 411.3.2.2. The time stated in Table 41.1 of the standard is also applicable to IT systems with a distributed or non-distributed neutral conductor or mid-point conductor. Chapter 4.2.3.1 of this book explains this in detail.

subsequent second insulation fault simultaneously on different active conductors and that way produces the required interrupting current. Therefore, the meaning of two insulation faults is neither two insulation faults in the same conductor, nor two insulation faults in the neutral conductor. Such double insulation faults do not pose a greater risk than a single insulation fault. On the contrary, the first fault according to IEC 60364-4-41, 411.6.3.1 with notes, is open to a different interpretation. The first insulation fault described there, in reality is a fault condition, with a certain insulation resistance not necessarily leading to automatic disconnection. This insulation resistance is the total value of the strand insulation resistance of all active conductors to earth or earthed parts, and may have any value between zero and infinity.

The definition of the insulation resistance is defined in IEC 61557-8:

"**Insulation resistance ($R_F$)**: Resistance in the system being monitored, including the resistance of all the connected appliances to earth."

The task of the insulation monitoring device is, to record and report the fault condition as described above, and to indicate it in a suitable manner.

Suitable insulation-fault-location-devices may be applied, in order to locate and eliminate the first insulation fault with the shortest practicable delay.

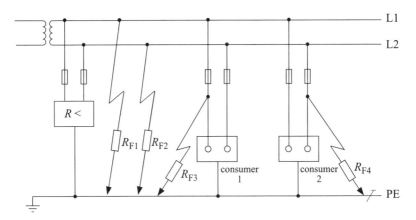

**Figure 13.7** Insulation fault constellations in an IT system
$R_{F1}$     "insulation fault condition" = 1. insulation fault
$R_{F2}$     "insulation fault condition" = 2. insulation fault
$R_{F3}$     "first insulation fault"         = 1. short-circuit to exposed-conductive-part or low ohmic insulation fault
$R_{F4}$     "second insulation fault"    = 2. short-circuit to exposed-conductive-part or low ohmic insulation fault

### 13.4.1 Fault Constellations in a. c. IT Systems

**Figure 13.7** shows the fault variations in an a. c. IT system. The fault constellations of three-phase or d. c. IT systems were omitted to keep the diagram clear.

The probable fault constellations according to figure 13.7 are described as follows:

An insulation monitoring device records and indicates the sum of all insulation faults occurring in an IT system ($R_{F1}$ and $R_{F2}$) and low ohmic insulation faults ($R_{F3}$ and $R_{F4}$); mere subsequent short-circuit to exposed-conductive-parts or low ohmic insulation faults on an already existing earth-connection on the same conductor do not lead to a change of the measuring results on the insulation monitoring device. Just-in-time fault clearance of the reported insulation drop serves as preventive maintenance of the system. The system however must be such, that after the event of a first low-resistance short-circuit to exposed-conductive-parts and low ohmic insulation faults ($R_{F3}$ and $R_{F4}$), the disconnection of the power supply in the event of a second short-circuit to exposed-conductive-parts or low ohmic insulation fault ($R_{F3}$ and $R_{F4}$) on another active conductor must be fulfilled.

## Literature

[13.1] Kaul, K.-H.: Ableitströme im IT-Netz. Bender-Information. Grünberg, 1990

[13.2] Herhahn, A.; Winkler, A.: Elektroinstallation nach VDE 0100. Würzburg: Vogel-Verlag, 1984

[13.3] Hofheinz, W.; Kaul, K.-H.; Schumann, W.: Überspannungen in Wechsel- und Drehstrom-IT-Systemen. etz Elektrotech. Z. (1998) H. 16, S. 20–23

[13.4] IEC 60364-4-41 Teil 413.1.5. Berlin: VDE VERLAG

[13.5] DIN VDE 0100-410: Schutzmaßnahmen; Schutz gegen gefährliche Körperströme. Berlin: VDE VERLAG

[13.6] Hofheinz, W.: Fault Current Monitoring in Electrical Installations, Berlin und Offenbach: VDE VERLAG, 2004

[13.7] Kindt, A: Berechnung von transienten Überspannungen in Niederspannungsnetzen mit IT-Systemen. Diplomarbeit, TU Dresden, 1995

[13.8] Hofheinz, W.; Kaul, K.; Schumann, W.: Überspannungen in Wechsel- und Drehstrom-IT-Systemen. etz Elektrotech. Z. 16 (1998) S. 20–23

[13.9] IEC GUIDE 104: 1997-08, The preparation of safety publications and the use basic safety publications and group publications

# 14 Standard References to IT Systems

This chapter lists the current IEC standards regarding insulation monitoring. The respective clauses are quoted in context. Standard drafts (CDs or CDVs) are considered.

## 14.1 IEC 60364-4-41:2005-12
### Low-voltage electrical installations – Part 4-41: Protection for safety – Protection against electric shock

*411.6 IT systems*

*411.6.1 to 411.6.4*

- See chapter 4.5 of this book for the details to these clauses

## 14.2 IEC 60364-4-43:2005-07, 64/1557/CDV, 64/1546/RVC, Low-voltage electrical installations – Part 4-43: Protection for safety – Protection against overcurrent

For more information on this subject see Chapter 5.4 of this book.

## 14.3 IEC 60364-5-53 Ed. 4/CDV:2005-11, IEC 64/1516/CDV Low-voltage electrical installations – Part 5-53: Selection and erection of electrical equipment – Protection, isolation, switching, control and monitoring

*531.1 IT systems*

For more information on this subject see Chapter 10.4 of this book.

## 14.4 IEC 60364-6:2005, Low-voltage electrical installations – Part 6: Verification

*61 Initial verification*

*61.3.6.1 General*

For more information on this standard see Chapter 5.3 of this book.

## 14.5 IEC 60364-7-710:2002-11 Electrical installations of buildings – Part 7-710: Requirements for special installations or locations – Medical locations

*710.413.1.5 Medical IT system*

Note 1: In the United States such a system is identified as an "Isolated Power System".

In group 2 medical locations, the medical IT system shall be used for circuits supplying medical electrical equipment and systems intended for life support, surgical applications and other electrical equipment located in the "patient environment", excluding equipment listed in 713.413.1.3.

For each group of rooms serving the same function, at least one separate medical IT system is necessary. The medical IT system shall be equipped with an insulation monitoring device in accordance with IEC 61557-8 with the following specific requirements:
- the a. c. internal impedance shall be at least 100 k$\Omega$
- the test voltage shall not be greater than 25 V d. c.
- the injected current, even under fault conditions, shall not be greater than 1 mA peak
- indication shall take place at the latest when the insulation resistance has decreased to 50 k$\Omega$. A test device shall be provided

Note 2: In Germany, an indication is required if the earth or wiring connection is lost.

Note 3: The necessary additional requirements on IMDs given above are at this time not covered in the equipment standard IEC 61557-8. They will be removed from this publication as soon as they have been treated in the relevant equipment standard.

For each medical IT system, an acoustic and visual alarm system incorporating the following components shall be arranged at a suitable place so that it can be permanently monitored (audible and visual signals) by the medical staff:

- *a green signal lamp to indicate normal operation;*
- *a yellow signal lamp which lights when the minimum value set for the insulation resistance is reached. It shall not be possible for this light to be cancelled or disconnected.*

### 710.512.1.1 Transformers for medical IT systems

### 710.512.1.6 Medical IT systems for group 2 medical locations

Transformers shall be in accordance with IEC 61558-2-15, with the following additional requirements:

The leakage current of the output winding to earth and the leakage current of the enclosure, when measured in no-load condition and the transformer supplied at rated voltage and rated frequency, shall not exceed 0,5 mA.

Single-phase transformers shall be used to form the medical IT systems for portable and fixed equipment and the rated output shall not be less than 0,5 kVA and shall not exceed 10 kVA.

If the supply of three-phase loads via an IT system is also required, a separate three-phase transformer shall be provided for this purpose with output line-to-line voltage not exceeding 250 V.

### 710.53.1 Protection of wiring systems in medical locations of group 2

Overcurrent protection against short-circuit and overload current is necessary for each final circuit. Overload current protection is not allowed in the feeder circuits upstream and downstream of the transformer of medical IT-system. Fuses may be used for short-circuit protection.

### 710.55.3 Socket-outlet circuits in the medical IT system for medical locations of group 2

At each patient's place of treatment, e.g. bed-heads, the configuration of socket-outlets shall be
as follows:
- either a minimum of two separate circuits feeding socket-outlets shall be installed or
- each socket-outlet shall be individually protected against overcurrent.
  Where circuits are supplied from other systems (TN-S or TT systems) in the same medical location, socket-outlets connected to the medical IT system shall either:
- be of such construction that prevents their use in other systems, or
- be clearly and permanently marked

### 710.61 Initial verification

The tests specified below under items a) to e) in addition to the requirements of IEC 60364-6-61, shall be carried out, both prior to commissioning and after alterations or repairs and before re-commissioning.

a) *Functional test of insulation monitoring devices of medical IT systems and acoustical/visual alarm systems.*

Note of the author:

More information on the medical IT system may be found in:

Wolfgang Hofheinz, Elektrische Sicherheit in medizinisch genutzten Bereichen, VDE VERLAG GmbH, Berlin und Offenbach, 2005 (not yet published in English)

## 14.6 IEC 60092-507, Ed. 2, IEC 18/1017/CDV:2005, Electrical installations in ships – Part 507: Small vessels

*5.2.3 Earthing of non-current-carrying parts in a.c. systems*

*In steel hulled vessels the presence of a large comparative cross section area of metal for earth return paths enables a simple method of earthing and bonding (TT and IT systems) in which non current carrying parts can be directly bonded to the vessel's hull. A vessel with non-metallic-hull requires to be provided with a protective conductor (which may be separate from the neutral conductor(TN-S) or (TN-C)).*

*6.4 Non-neutral earthed system (IT type system)*

*IT type systems are permissible when continuity of service is requested otherwise, TN systems is acceptable. In IT systems the vessel's phase conductor(s) is insulated from earth and the star point either isolated from earth or deliberately connected to earth through a sufficiently high impedance.*

*In IT type systems, it is permissible for a single fault between a live part and an exposed conductive part to occur without automatic disconnection, provided that earth monitoring or permanent insulation controller is fitted. A second fault shall result in automatic disconnection.*

*A prospective touch voltage exceeding 50 V a.c. shall not persist for a time sufficient to cause a risk of harmful physiological effect in a person.*

*In IT type systems, the protective arrangements particularly to locations in confined or*
*exceptionally damp spaces where particular risk due to conductivity may exist shall be*
*i) an overcurrent protective device and*
*ii) a residual-current protective device with sensitivity of 30 mA maximum for final circuits to locations where there is an increased risk of personal contact with live conductive parts.*

## 12.3.2 Non-earthed system (IT type system)

*Insulation-monitoring devices shall be fitted to give warning of earth faults and assist in the location of the fault.*

### A.1.3 Information and instructions for connecting a vessel to a shore supply

*General*

*b) The supply voltage at this marina is supplied by socket-outlets complying with IEC 60309-2, position 6 h or position 9 h for three-phase in case of an IT system).*


Note of the author:

More information on IT system application on board ships see Chapter 7.2 of this book.

# 15 Definitions for Insulation Monitoring

## 15.1 Definitions in accordance with IEC 61557-8:2006[1]

| German | English | French |
|---|---|---|
| **Fremdgleichspannung $U_{fg}$**<br>Die Fremdgleichspannung ist die Gleichspannung in Wechselspannungsnetzen, die zwischen Netz und Erde auftritt (abgeleitet aus Gleichspannungsteilen) | **extraneous d.c. voltage $U_{fg}$**<br>d.c. voltage occurring in a.c. systems between the a.c. conductors and earth (derived from d.c. parts) | **Contrôleurs permanents d'isolement**<br>De par leur principe même de fonctionnement, les contrôleurs permanents d'isolement doivent être en mesure de signaler des détériorations tant symétriques qu'asymétriques de l'isolement. (provoqué par des parties en tension continue) |
| **Isolationswiderstand $R_F$**<br>Der Isolationswiderstand ist der Wirkwiderstand des überwachten Netzes einschließlich der Wirkwiderstände aller daran angeschlossenen Betriebsmittel gegen Erde | **insulation resistance $R_F$**<br>resistance in the system being monitored, including the resistance of all the connected appliances to earth | **Résistance d'isolement $R_F$**<br>Il s'agit de la résistance effective par rapport à la terre du réseau surveillé et des matériels que y sont connectés |
| **Sollansprechwert $R_{an}$**<br>Der Sollansprechwert ist der am Gerät fest eingestellte oder einstellbare Wert des Isolationswiderstands, dessen Unterschreitung überwacht wird | **specified response value $R_{an}$**<br>value of the insulation resistance, permanently set or adjustable on the device and monitored if the insulation resistance falls below this limit | **Valeur de seuil de référence $R_{an}$**<br>Il s'agit de la valeur de la résistance d'isolement qui est préréglée ou réglable sur l'appareil et dont le dépassement est surveillé |
| **Ansprechwert $R_a$**<br>Der Ansprechwert ist der Wert des Isolationswiderstands, bei dem das Gerät unter festgelegten Bedingungen anspricht | **response value $R_a$**<br>value of the insulation resistance at which the device responds under specified conditions | **Valeur de seuil $R_a$**<br>Il s'agit de la valeur de la résistance d'isolement à laquelle l'appareil réagit dans des conditions données |

---

[1] Series IEC 61557 parts 1 to 8 are currently being revised, the 2nd editions are expected to be published in 2006.

| German | English | French |
|---|---|---|
| **Relative (prozentuale) Messunsicherheit** $A[\%]$<br>Die Ansprechabweichung ist der Ansprechwert, vermindert um den Sollansprechwert, geteilt durch den Sollansprechwert, multipliziert mit hundert, angegeben in Prozent:<br><br>$$A = \frac{R_a - R_{an}}{R_{an}} \cdot 100\,\%$$ | **relative (percentage) uncertainty** $A[\%]$<br>response value minus the specified response value, divided by the specified response value, multiplied by 100 and stated as a percentage:<br><br>$$A = \frac{R_a - R_{an}}{R_{an}} \cdot 100\,\%$$ | **Incertitude relative** $A[\%]$<br>Il s'agit de la valeur de seuil de laquelle est soustraite la valeur de seuil de référence, divisée par la valeur de seuil de référence, multipliée par cent et donnée en pourcentage:<br><br>$$A = \frac{R_a - R_{an}}{R_{an}} \cdot 100\,\%$$ |
| **Netzableitkapazität** $C_e$<br>Die Netzableitkapazität ist der maximal zulässige Wert der Gesamtkapazität des zu überwachenden Netzes einschließlich aller angeschlossenen Betriebsmittel gegen Erde, bis zu dem ein Isolationsüberwachungsgerät bestimmungsgemäß arbeiten kann | **system leakage capacitance** $C_e$<br>maximum permissible value of the total capacitance to earth of the system to be monitored, including any connected appliances, up to which value the insulation monitoring device can work as specified | **Capacité de fuite au réseau** $C_e$<br>Il s'agit de la valeur maximale admissible de la capacité totale par rapport à la terre du réseau à surveiller et de tous les matériels connectés, jusqu'à laquelle l'appareil peut travailler conformément aux prescriptions |
| **Bemessungsspannung**<br>Spannung für die ein Relais Kontakt bemessen ist, unter bestimmten Bedingungen zu öffnen oder zu schließen | **rated contact voltage**<br>voltage for which a relay contact is rated to open and close under specified conditions | **Tension assignée**<br>Tension pour laquelle, dans des conditions données, un contact de relais est assigné à ouvrir ou à fermer |
| **Ansprechzeit** $t_{an}$<br>Die Ansprechzeit ist die Zeit, die ein Isolationsüberwachungsgerät zum Ansprechen unter vorgegebenen Bedingungen benötigt | **response time** $t_{an}$<br>time required by an insulation monitoring device to respond under the conditions specified | **Temps de réponse** $t_{an}$<br>Il s'agit du temps nécessaire à un contrôleur permanent d'isolement pour réagir dans des conditions données |
| **Messspannung** $U_m$<br>Die Messspannung ist die Spannung, die während der Messung an den Messanschlüssen vorhanden ist<br>*Anmerkung: Im fehlerfreien, spannungslosen Netz ist dies die Spannung, die zwischen den Anschlussklemmen am zu überwachenden Netz und den Schutzleiterklemmen anliegt* | **measuring voltage** $U_m$<br>voltage present at the measuring terminals during the measurement<br>*Note: Additionally to the definition in IEC 61557-1, the measuring voltage ($U_m$) is the voltage present in a fault-free and de-energized system between the terminals of the system to be monitored and the terminals of the protective conductor* | **Tension de mesure** $U_m$<br>Il s'agit de la tension qui existe aux bornes de mesure pendant les essais<br>*Note: Dans un réseau hors tension et dépourvu de défaut, il s'agit de la tension qui se trouve entre les bornes de raccordement situées sur le réseau à surveiller et les bornes du conducteur de protection* |

| German | English | French |
|---|---|---|
| **Messstrom $I_m$** Der Messstrom ist der maximale Strom, der aus der Messspannungsquelle, begrenzt durch den Innenwiderstand $R_i$ des Isolationsüberwachungsgeräts, zwischen Netz und Erde fließen kann | **measuring current $I_m$** maximum current that can flow between the system and earth, limited by the internal resistance $R_i$ from the measuring voltage source of the insulation monitoring device | **Courant de mesure $I_m$** Il s'agit du courant maximal qui s'écoule de la source de la tension de mesure et qui peut circuler entre le réseau et la terre. Il est limité par la résistance interne $R_i$ du contrôleur permanent d'isolement |
| **Wechselstrom-Innenwiderstand $Z_i$** Der Wechselstrom-Innenwiderstand ist die Gesamtimpedanz des Isolationsüberwachungsgeräts zwischen Netz- und Erdanschlüssen bei Nennfrequenz | **internal impedance $Z_i$** total impedance of the insulation monitoring device between the terminals to the system being monitored and earth, measured at the nominal frequency | **Résistance interne du courant alternatif $Z_i$** Il s'agit de l'impédance totale du contrôleur permanent d'isolement entre les bornes du réseau et de la terre en fréquence nominale |
| **Gleichstrom-Innenwiderstand $R_i$** Der Gleichstrom-Innenwiderstand ist der Wirkwiderstand des Isolationsüberwachungsgeräts zwischen Netz- und Erdanschlüssen | **internal d. c. resistance $R_i$** resistance of the insulation monitoring device between the terminals to the system being monitored and earth | **Résistance interne du courant continu $R_i$** Il s'agit de la résistance effective du contrôleur permanent d'isolement entre des bornes du réseau et de la terre |
| **Funktionserdverbindung FE** Elektrische Verbindung zwischen dem RCM und Erdpotential ist vorgesehen, um sicherzustellen, dass: <br>• der Bezugspunkt für RCMs mit Selektivitätsfunktionen vorhanden ist und oder <br>• im Falle des Verlusts des Neutralleiters auf der Netzseite ein korrektes Ansprechen durchgeführt werden kann | **functional earthing FE** earthing a point or points in a system or in an installation or in equipment for purposes other than electrical safety *Note: For insulation monitoring devices this is the measuring connection to earth* | **Liaison à la terre fonctionnelle FE** Liaison électrique qui est prévue entre le RCM et le potentiel de terre afin de garantir <br>• que le point de référence existe pour les RCM dotés de fonctions de sélectivité et/ou <br>• qu'une réaction correcte puisse avoir lieu dans le cas de la perte du conducteur neutre du côté du réseau |

| Other: | | |
|---|---|---|
| **Isolationsüberwachungsgerät** Isolationsüberwachungsgeräte müssen von dem ihnen vorgegebenen Messprinzip in der Lage sein, sowohl symmetrische als auch unsymmetrische Isolationsverschlechterungen zu melden | **Insulation monitoring devices** Insulation monitoring devices shall be capable to monitor symmetrical as well as asymmetrical insulation deteriorations according to the stipulated measuring principle | **Contrôleurs permanents d'isolement** De par leur principe même de fonctionnement, les contrôleurs permanents d'isolement doivent être en mesure de signaler des détériorations tant symétriques qu'asymétriques de l'isolement |
| **Symmetrische Isolationsverschlechterung** Eine symmetrische Isolationsverschlechterung liegt dann vor, wenn sich der Isolationswiderstand aller Leiter des zu überwachenden Netzes (annähernd) gleichmäßig verringert | **Symmetrical Insulation deterioration** A symmetrical insulation deterioration occurs when the insulation resistance of all conductors in the system to be monitored decreases (approximately) similarly | **Détérioration symétrique de l'isolement** Une détérioration de l'isolement existe lorsque la résistance d'isolement de l'ensemble des conducteurs du réseau à surveiller décroit régulièrement |
| **Unsymmetrische Isolationsverschlechterung** Eine unsymmetrische Isolationsverschlechterung liegt dann vor, wenn sich der Isolationswiderstand, z. B. eines Leiters, wesentlich stärker verringert als der der (des) übrigen Leiter(s) | **Asymmetrical Insulation deterioration** An asymmetrical insulation deterioration occurs when the insulation resistance of, for example one conductor, decreases (substantially) more than that of the other conductor(s) | **Détérioration asymétrique de l'isolement** Une détérioration asymétrique de l'isolement existe lorsque la résistance d'isolement d'un conducteur par exemple, décroit davantage que celle de l'autre ou des autres conducteur(s) |

# 16 Abbreviations

| | |
|---|---|
| ANSI | American National Standards Institute, New York, USA |
| | Amerikanische nationale Normeninstitution |
| ASTM | American Society for Testing Materials, Philadelphia, USA |
| BS | British Standard |
| | Britische Norm |
| BSI | British Standards Institution, London, UK |
| | Britisches Normeninstitut |
| CEI | Commission Électrotechnique Internationale |
| | Internationale Elektrotechnische Kommission (IEC) |
| CENELEC | European Committee for Electrotechnical Standardisation, Brüssel |
| | Europäisches Komitee für Elektrotechnische Normung |
| DKE | German Commission for Electrical, Electronic & Information Technologies of DIN and VDE, Frankfurt a. M. |
| | Deutsche Kommission Elektrotechnik Elektronik Informationstechnik im DIN und VDE, Frankfurt a. M. |
| DIN | German Standards Institute |
| | Deutsches Institut für Normung, Berlin |
| EN | European Standard |
| | Europäische Norm |
| EWG | European Community |
| | Europäische Wirtschaftsgemeinschaft |
| EU | European Union |
| | Europäische Union |
| HD | Harmonization Document (CENELEC) |
| | Harmonisierungsdokument |
| IEC | International Electrotechnical Commission, Geneva |
| | Internationale Elektrotechnische Kommission |

| | |
|---|---|
| IEE | Institution of Electrical Engineers (UK), London |
| IEEE | Institute of Electrical and Electronics Engineers, New York, USA |
| K | Committee of DKE |
| | Komitee in der DKE Deutsche Kommission Elektrotechnik Elektronik Informationstechnik, z. B. K 221 |
| NFPA | National Fire Protection Association, Quincy, USA |
| NEC | National Electrical Code, USA |
| UTE | L'Union Technique de l'Électricité, Paris, Frankreich |
| UK | Subcommittee of DKE |
| | Unterkomitee in der DKE Deutsche Kommission Elektrotechnik Elektronik Informationstechnik, z. B. UK 964.1 |
| UL | Underwriters Laboratories Inc., USA |
| VDE | Association for Electrical, Electronic & Information Technology, Frankfurt a. M. |
| | Verband Elektrotechnik Elektronik Informationstechnik e. V.. Frankfurt a. M. |

# 17 List of Referenced IEC Standards

All IEC Standards are available in both English, French, and some in Spanish.

| International Standard | Year of Publication | Title | Corresponding German Standard | |
|---|---|---|---|---|
| | | | Title | Publication date |
| IEC 60038 | Consolidated Edition 6.2 (incl. am1+am2) (2002-07) | IEC STANDARD VOLTAGES | IEC-Normspannungen | DIN IEC 60038 (VDE 0175): 2002-11 |
| IEC 60077-1 | Edition 1.0 (1999-10) | Railway applications – Electric equipment for rolling stock – Part 1: General service conditions and general rules | Bahnanwendungen – Elektrische Betriebsmittel auf Bahnfahrzeugen – Teil 1: Allgemeine Betriebsbedingungen und allgemeine Regeln (IEC 60077-1:1999, modifiziert); Deutsche Fassung EN 60077-1:2002 | DIN EN 60077-1 (VDE 0115-460-1): 2003-04 |
| IEC 60077-2 | Edition 1.0 (1999-03) | Railway applications – Electric equipment for rolling stock – Part 2: Electrotechnical components – General rules | Bahnanwendungen – Elektrische Betriebsmittel auf Bahnfahrzeugen – Teil 2: Elektrotechnische Bauteile; Allgemeine Regeln (IEC 60077-2:1999, modifiziert); Deutsche Fassung EN 60077-2:2002 | DIN EN 60077-2 (VDE 0115-460-2): 2003-04 |
| IEC 60077-3 | Edition 1.0 (2001-12) | Railway applications – Electric equipment for rolling stock – Part 3: Electrotechnical components – Rules for d. c. circuit-breakers | Bahnanwendungen – Elektrische Betriebsmittel auf Bahnfahrzeugen – Teil 3: Elektrotechnische Bauteile; Regeln für DC-Leistungsschalter (IEC 60077-3:2001); Deutsche Fassung EN 60077-3:2002 | DIN EN 60077-3 (VDE 0115-460-3): 2003-04 |
| IEC 60077-4 | Edition 1.0 (2003-02) | Railway applications – Electric equipment for rolling stock – Part 4: Electrotechnical components – Rules for AC circuit-breakers | Bahnanwendungen – Elektrische Geräte auf Bahnfahrzeugen – Teil 4: Elektrotechnische Bauteile – Regeln für AC-Leistungsschalter (IEC 60077-4:2003); Deutsche Fassung EN 60077-4:2003 | DIN EN 60077-4 (VDE 0115-460-4): 2004-01 |
| IEC 60077-5 | Edition 1.0 (2003-07) | Railway applications – Electric equipment for rolling stock – Part 5: Electrotechnical components – Rules for HV fuses | Bahnanwendungen – Elektrische Betriebsmittel auf Bahnfahrzeugen – Teil 5: Elektrotechnische Bauteile – Regeln für Hochspannungssicherungen (IEC 60077-5:2003); Deutsche Fassung EN 60077-5:2003 | DIN EN 60077-5 (VDE 0115-460-5): 2004-07 |

| International Standard | Year of Publication | Title | Corresponding German Standard | |
|---|---|---|---|---|
| | | | Title | Publication date |
| IEC 60092-101 | Consolidated Edition 4.1 (incl. am1) (2002-08) | Electrical installations in ships – Part 101: Definitions and general requirements | Elektrische Anlagen auf Schiffen –Teil 507: Yachten ( IEC 60092-507: 2000); Deutsche Fassung EN 60092-507:2000 | DIN EN 60092-507 (VDE 0129-507): 2001-11 |
| IEC 60092-201 | Edition 4.0 (1994-08) | Electrical installations in ships – Part 201: System design – General | | |
| IEC 60092-502 | Edition 5.0 (1999-02) | Electrical installations in ships – Part 502: Tankers – Special features | | |
| IEC 60092-503 | Edition 1.0 (1975-01) | Electrical installations in ships. Part 503: Special features – A. C. supply systems with voltages in the range above 1 kV up to and including 11 kV | | |
| IEC 60092-507 | Edition 1.0 (2000-02) | Electrical installations in ships – Part 507: Pleasure craft | | |
| IEC 60309-2 | Consolidated Edition 4.1 (incl. am1) (2005-12) | Plugs, socket-outlets and couplers for industrial purposes – Part 2: Dimensional interchangeability requirements for pin and contact-tube accessories | Stecker, Steckdosen und Kupplungen für industrielle Anwendungen –Teil 2: Anforderungen und Hauptmaße für die Austauschbarkeit von Stift-und Buchsensteckvorrichtungen (IEC 60309-2:1999); Deutsche Fassung EN 60309-2:1999 | DIN EN 60309-2 (VDE 0623-20): 2000-05 DIN EN 60309-2/A11 (VDE 0623-20/A1): 2004-1 |
| IEC 60310 | Edition 3.0(2004-02) | Railway applications – Traction transformers and inductors on board rolling stock | Bahnanwendungen – Transformatoren und Drosselspulen auf Bahnfahrzeugen (IEC 60310:2004); Deutsche Fassung EN 60310:2004 | DIN EN 60310 (VDE 0115-420): 2005-01 |
| IEC 60335-1 | 2004-07 | Household and similar electrical appliances – Safety – Part 1: General requirements | Sicherheit elektrischer Geräte für den Hausgebrauch und ähnliche Zwecke – Teil 1: Allgemeine Anforderungen | DIN EN 60335-1 (VDE 0700-1): 2003-07 |

| International Standard | Year of Publication | Title | Corresponding German Standard | |
|---|---|---|---|---|
| | | | Title | Publication date |
| IEC 60349-1 | Consolidated Edition 1.1 (incl. am1) (2002-10) | Electric traction - Rotating electrical machines for rail and road vehicles – Part 1: Machines other than electronic convertor-fed alternating current motors | Elektrische Zugförderung – Drehende elektrische Maschinen für Bahn- und Straßenfahrzeuge – Teil 1: Elektrische Maschinen, ausgenommen umrichtergespeiste Wechselstrommotoren (IEC 60349-1:1999 + A1:2002); Deutsche Fassung EN 60349-1:2000 + A1:2002 | DIN EN 60349 (VDE 0115-400-1): 2003-07 |
| IEC 60364-1 | Edition 5.0 (2005-11) | Low-voltage electrical installations – Part 1: Fundamental principles, assessment of general characteristics, definitions | Errichten von Niederspannungsanlagen – Teil 100: Allgemeine Grundsätze, Bestimmungen allgemeiner Merkmale, Begriffe (IEC 64/1295/CD:2003) | DIN VDE 0100-100 (VDE 0100-100): 2002-08 E DIN IEC 60364-1 (VDE 0100-100): 2003-08 |
| IEC 60364-4-41 | Edition 5.0 (2005-12) | Low-voltage electrical installations – Part 4-41: Protection for safety – Protection against electric shock | Errichten von Niederspannungsanlagen – Teil 4-41: Schutzmaßnahmen; Schutz gegen elektrischen Schlag (IEC 64/1272/CDV:2002) | DIN VDE 0100-410 (VDE 0100-410): 1997-01 E DIN VDE 0100-410 (VDE 0100-410):2003-04/A1: 2003-06 |
| IEC 60364-5-53 | Consolidated Edition 3.1 (incl. am1) (2002-06) | Electrical installations of buildings – Part 5-53: Selection and erection of electrical equipment – Isolation, switching and control | Errichten von Niederspannungsanlagen – Teil 5-53: Auswahl und Errichtung elektrischer Betriebsmittel – Trennen, Schalten und Steuern (IEC 64/1486/CD:2005) | DIN VDE 0100-530 (VDE 0100-530): 2005-06 E DIN IEC 60364-5-53 (VDE 0100-530):2005-12 |
| IEC 60364-5-54 | Edition 2.0 (2002-06) | Electrical installations of buildings – Part 5-54: Selection and erection of electrical equipment – Earthing arrangements, protective conductors and protective bonding conductors | Errichten von Niederspannungsanlagen – Teil 5-54: Auswahl und Errichtung elektrischer Betriebsmittel; Erdungsanlagen, Schutzleiter und Potentialausgleich (IEC 64/1134/CD:2000) | DIN VDE 0100-540 (VDE 0100-540): 1991-11 E DIN IEC 64/1134/CD (VDE 0100-540):2000-11 |

| International Standard | Year of Publication | Title | Corresponding German Standard Title | Publication date |
|---|---|---|---|---|
| IEC 60364-5-55 | Consolidated Edition 1.1 (incl. am1) (2002-05) | Electrical installations of buildings – Part 5-55: Selection and erection of electrical equipment – Other equipment | Elektrische Anlagen von Gebäuden – Teil 5: Auswahl und Errichtung elektrischer Betriebsmittel; Kapitel 55: Andere Betriebsmittel; Hauptabschnitt 551: Niederspannungs-Stromversorgungsanlagen (IEC 60364-5-551:1994); Deutsche Fassung HD 384.5.551 S1:1997 | DIN VDE 0100-551 (VDE 0100-551): 1997-08 |
| IEC 60364-6 | Edition 1.0 (2006-02) | Low-voltage electrical installations – Part 6: Verification | Errichten von Niederspannungsanlagen – Teil 6-61: Prüfungen – Erstprüfungen (IEC 60364-6-61:1986 + A1:1993 + A2: 1997, modifiziert); Deutsche Fassung HD 384.6.61 S2:2003 | DIN VDE 0100-610 (VDE 0100-610): 2004-04 |
| IEC 60364-7-710 | Edition 1.0 (2002-11) | Electrical installations of buildings – Part 7-710: Requirements for special installations or locations – Medical locations | Errichten von Niederspannungsanlagen – Anforderungen für Betriebsstätten, Räume und Anlagen besonderer Art – Teil 710: Medizinisch genutzte Bereiche | DIN EN 60364-710 (VDE 0100-710): 2002-11 |
| IEC/TS 60479-1 | Edition 4.0 (2005-07) | Effects of current on human beings and livestock – Part 1: General aspects | Wirkungen des elektrischen Stromes auf Menschen und Nutztiere – Allgemeine Aspekte Identisch mit IEC-Report 479-1:1994 | DIN V VDE V0140-479 (VDE V 0140-479): 1996-02 |
| IEC/TR 60479-2 | Edition 2.0 (1987-03) | Effects of current passing through the human body. Part 2: Special aspects – Chapter 4: Effects of alternating current with frequencies above 100 Hz – Chapter 5: Effects of special waveforms of current – Chapter 6: Effects of unidirectional single impulse currents of short duration | | |
| IEC/TR 60479-3 | Edition 1.0 (1998-09) | Effects of current on human beings and livestock – Part 3: Effects of currents passing through the body of livestock | Wirkungen des elektrischen Stromes auf Menschen und Nutztiere – Teil 3: Wirkungen von Strömen durch den Körper von Nutztieren; Identisch mit IEC-Report 60479-3: 1998 | DIN V VDE V0140-479 (VDE V 0140-479): 2001-04 |

| International Standard | Year of Publication | Title | Corresponding German Standard | |
|---|---|---|---|---|
| | | | Title | Publication date |
| IEC/TR 60479-4 | Edition 1.0 (2004-07) | Effects of current on human beings and livestock – Part 4: Effects of lightning strokes on human beings and livestock | Wirkungen des Stromes auf Menschen und Nutztiere – Teil 4: Wirkungen von Blitzschlägen auf Menschen und Nutztiere (IEC/ TR 60479-4:2004) | DIN V VDE V 0140-479 (VDE V 0140-479): 2005-10 |
| IEC 60721-3-1 | Edition 2.0 (1997-02) | Classification of environmental conditions – Part 3 Classification of groups of environmental parameters and their severities – Section 1: Storage | | |
| IEC 60721-3-2 | Edition 2.0 (1997-03) | Classification of environmental conditions – Part 3: Classification of groups of environmental parameters and their severities – Section 2: Transportation | | |
| IEC 60721-3-3 | Consolidated Edition 2.2 (incl. am1+am2) (2002-10) | Classification of environmental conditions – Part 3-3: Classification of groups of environmental parameters and their severities – Stationary use at weatherprotected locations | | |
| IEC 60850 | Edition 2.0 (2000-08) | Railway applications – Supply voltages of traction systems | | |
| IEC 60950-1 | Edition 2.0 (2005-12) | Information technology equipment – Safety – Part 1: General requirements | Einrichtungen der Informationstechnik – Sicherheit – Teil 1: Allgemeine Anforderungen | DIN EN 60950-1 (VDE 0805-1): 2003-03 |
| IEC 61008-1 | Consolidated Edition 2.2 (incl. am1 + am2) (2006-06) | Residual current operated circuit-breakers without integral overcurrent protection for household and similar uses (RCCBs) – Part 1: General rules | Fehlerstrom-/ Differenzstrom-Schutzschalter ohne eingebauten Überstromschutz ( RCCBs) für Hausinstallationen und für ähnliche Anwendungen – Teil 1: Allgemeine Anforderungen ( IEC 61008-1:1996 + A1:2002, modifiziert); Deutsche Fassung EN 61008-1:2004 | DIN EN 61008-1 (VDE 0664-10): 2005-06 |

| International Standard | Year of Publication | Title | Corresponding German Standard | |
|---|---|---|---|---|
| | | | Title | Publication date |
| IEC 61008-2-1 | Edition 1.0 (1990-12) | Residual current operated circuit-breakers without integral overcurrent protection for household and similar uses (RCCB's) – Part 2-1: Applicability of the general rules to RCCB's functionally independent of line voltage | Fehlerstrom-/Differenzstrom-Schutzschalter ohne eingebauten Überstromschutz (RCCBs) für Hausinstallationen und für ähnliche Anwendungen – Teil 2-1: Anwendung der allgemeinen Anforderungen auf netzspannungsunabhängige RCCBs | DIN EN 61008-2-1 (VDE 0664-11): 1999-12 |
| IEC 61009-1 | Consolidated Edition 2.2 (incl. am1 + am2) (2006-06) | Residual current operated circuit-breakers with integral overcurrent protection for household and similar uses (RCBOs) – Part 1: General rules | Fehlerstrom-/ Differenzstrom-Schutzschalter mit eingebautem Überstromschutz (RCBOs) für Hausinstallationen und für ähnliche Anwendungen – Teil 1: Allgemeine Anforderungen | DIN EN 61009 (VDE 0664-20): 2001-09 |
| IEC 61010-1 | Edition 2.0 (2001-02) | Safety requirements for electrical equipment for measurement, control, and laboratory use – Part 1: General requirements | Sicherheitsbestimmungen für elektrische Mess-, Steuer-, Regel- und Laborgeräte – Teil 1: Allgemeine Anforderungen (IEC 61010-1:2001); Deutsche Fassung EN 61010-1:2001 | DIN EN 61010-1 (VDE 0411-1): 2002-08, E: 2004-03 |
| IEC 61140 | Edition 3.0 (2001-10) am1 (2004-10) | Protection against electric shock – Common aspects for installation and equipment | Schutz gegen elektrischen Schlag – Gemeinsame Anforderungen für Anlagen und Betriebsmittel (IEC 61140:2001) Deutsche Fassung EN 61140:2002 | DIN EN 61140 (VDE 140 parte 1): 2003-08 |
| IEC/TR 61201 | Edition 1.0 (1992-09) | Extra-low voltage (ELV) – Limit values | | |
| IEC 61287-1 | Edition 2.0 (2005-09) | Railway applications – Power convertors installed on board rolling stock – Part 1: Characteristics and test methods | Bahnanwendungen – Stromrichter auf Bahnfahrzeugen – Teil 1: Eigenschaften und Prüfverfahren (IEC 9/742/CDV: 2003); Deutsche Fassung prEN 61287-1: 2003 | DIN EN 50207 (VDE 0115-410): 2001-03 E DIN EN 61287-1 (VDE 0115-410): 2003-07 |

| International Standard | Year of Publication | Title | Corresponding German Standard | |
|---|---|---|---|---|
| | | | Title | Publication date |
| IEC 61557-1 | Edition 1.0 (1997-02) | Electrical safety in low voltage distribution systems up to 1000 V a. c. and 1500 V d. c. – Equipment for testing, measuring or monitoring of protective measures – Part 1: General requirements | Elektrische Sicherheit in Niederspannungsnetzen bis AC 1000 V und DC 1500 V – Geräte zum Prüfen, Messen oder Überwachen von Schutzmaßnahmen – Teil 1: Allgemeine Anforderungen (IEC 61557-1:1997); Deutsche Fassung EN 61557-1:1997 | DIN EN 61557-1 (VDE 0413-1): 1998-05* |
| IEC 61557-2 | Edition 1.0 (1997-02) | Electrical safety in low voltage distribution systems up to 1000 V a. c. and 1500 V d. c. – Equipment for testing, measuring or monitoring of protective measures – Part 2: Insulation resistance | Elektrische Sicherheit in Niederspannungsnetzen bis AC 1000 V und DC 1500 V – Geräte zum Prüfen, Messen oder Überwachen von Schutzmaßnahmen – Teil 2: Isolationswiderstand (IEC 61557-2:1997); Deutsche Fassung EN 61557-2:1997 | DIN EN 61557-2 (VDE 0413-1): 1998-05* |
| IEC 61557-6 | Edition 1.0 (1997-02) | Electrical safety in low voltage distribution systems up to 1000 V a. c. and 1500 V d. c. – Equipment for testing, measuring or monitoring of protective measures – Part 6: Residual current devices (RCD) in TT and TN systems | Elektrische Sicherheit in Niederspannungsnetzen bis AC 1000 V und DC 1500 V – Geräte zum Prüfen, Messen oder Überwachen von Schutzmaßnahmen – Teil 6: Fehlerstrom-Schutzeinrichtungen (RCD) in TT-, TN-und IT-Netzen (IEC 61557-6:1997, modifiziert); Deutsche Fassung EN 61557-6:1998 | DIN EN 61557-6 (VDE 0413-1): 1999-05* |
| IEC 61557-8 | Edition 1.0 (1997-02) | Electrical safety in low voltage distribution systems up to 1000 V a. c. and 1500 V d. c. – Equipment for testing, measuring or monitoring of protective measures – Part 8: Insulation monitoring devices for IT systems | Geräte zum Prüfen, Messen oder Überwachen von Schutzmaßnahmen – Elektrische Sicherheit in Niederspannungsnetzen bis AC 1000 V und DC 1500 V Teil 8: Isolationsüberwachungsgeräte für IT-Netze (IEC 61557-8:1997) Deutsche Fassung EN 61557-8:1997 | DIN EN 61557-8 (VDE 0413-8): 1998-05* |

| International Standard | Year of Publication | Title | Corresponding German Standard | |
|---|---|---|---|---|
| | | | Title | Publication date |
| IEC 61557-9 | Edition 1.0 (1999-09) | Electrical safety in low voltage distribution systems up to 1 000 V a. c. and 1 500 V d.c. – Equipment for testing, measuring or monitoring of protective measures – Part 9: Equipment for insulation fault location in IT systems | Elektrische Sicherheit in Niederspannungsnetzen bis AC 1 kV und DC 1,5 kV – Geräte zum Prüfen, Messen oder Überwachen von Schutzmaßnahmen - Teil 9: Einrichtungen zur Isolationsfehlersuche in IT-Systemen (IEC 61557-9:1999) Deutsche Fassung EN 61557-9:1999 | DIN EN 61557-9 (VDE 0413-9): 2000-08 |
| IEC 61557-10 | Edition 1.0 (2000-12) | Electrical safety in low voltage distribution systems up to 1 000 V a. c. and 1 500 V d.c. – Equipment for testing, measuring or monitoring of protective measures – Part 10: Combined measuring equipment for testing, measuring or monitoring of protective measures | Elektrische Sicherheit in Niederspannungsnetzen bis AC 1 kV und DC 1,5 kV – Geräte zum Prüfen, Messen oder Überwachen von Schutzmaßnahmen - Teil 10: Kombinierte Messgeräte zum Prüfen, Messen oder Überwachen von Schutzmaßnahmen (IEC 61557-10:2000) Deutsche Fassung EN 61557-10:2001 | DIN EN 61557-10 (VDE 0413-10): 2001-12 |
| IEC 61558-2-15 | Edition 1.0 (1999-02) | Safety of power transformers, power supply units and similar – Part 2-15: Particular requirements for isolating transformers for the supply of medical locations | Sicherheit von Transformatoren, Netzgeräten und dergleichen – Teil 2-15: Besondere Anforderungen an Trenntransformatoren zur Versorgung medizinischer Räume (IEC 61558-2-15:1999, modifiziert); Deutsche Fassung EN 61558-2-15:2001 | DIN EN 61558-2-15 (VDE 0570-2-15): 2001-11 |
| IEC 61851-1 | Edition 1.0 (2001-01) | Electric vehicle conductive charging system – Part 1: General requirements | Konduktive Ladesysteme für Elektrofahrzeuge – Elektrische Ausrüstung von Elektro-Straßenfahrzeugen – Teil 1: Allgemeine Anforderungen | DIN EN 61851-1 (VDE 0122-1): 2001-11 |
| IEC 61851-21 | Edition 1.0 (2001-05) | Electric vehicle conductive charging system – Part 21: Electric vehicle requirements for conductive connection to an a.c./d.c. supply | Konduktive Ladesysteme für Elektrofahrzeuge – Elektrische Ausrüstung von Elektro-Straßenfahrzeugen – Teil 2-1: Anforderung eines Elektrofahrzeuges für konduktive Verbindung an a.c./d.c.-Versorgung | DIN EN 61851-21 (VDE 0122-2-1): 2002-10 |

| International Standard | Year of Publication | Title | Corresponding German Standard | |
|---|---|---|---|---|
| | | | Title | Publication date |
| IEC 61851-22 | Edition 1.0 (2001-05) | Electric vehicle conductive charging system – Part 22: AC electric vehicle charging station | Konduktive Ladesysteme für Elektrofahrzeuge – Elektrische Ausrüstung von Elektro-Straßenfahrzeugen – Teil 2-2: Wechselstrom-Ladestation für Elektrofahrzeuge | DIN EN 61851-22 (VDE 0122-2-2): 2002-10 |
| IEC 61991 | Edition 1.0 (2000-01) | Railway applications – Rolling stock – Protective provisions against electrical hazards | | |
| IEC 62020 | Consolidated Edition 1.1 (incl. am1) (2003-11) | Electrical accessories – Residual current monitors for household and similar uses (RCMs) | Elektrisches Installationsmaterial – Differenzstrom-Überwachungsgeräte für Hausinstallationen und ähnliche Verwendungen ( RCMs) (IEC 62020: 1998 + A1:2003, modifiziert); Deutsche Fassung EN 62020:1998 + A1:2005 | DIN EN 62020 (VDE 0663): 2005-11 |
| IEC 62236-5 | Edition 1.0 (2003-04) | Railway applications – Electromagnetic compatibility – Part 5: Emission and immunity of fixed power supply installations and apparatus | | |
| OTHER | | | Contact: | |
| IMO | 2002 edition | International Safety Management Code (ISM Code) and Guidelines on Implementation of the ISM Code | | |
| IEEE – Institute of Electrical and Electronics Engineers | 4th edition, June 1999 | IEEE VuSpec: Safety & Security Standards Series, Volume III: Electrical API RP 14F-1999, Recommended Practice for Design and Installation of Electrical Systems for Fixed and Floating Offshore Petroleum Facilities for Unclassified and Class I, Division 1 and Division 2 Locations | www.ieee.org | |

**Practical Information:**
To conduct your own research:

**IEC Central Office**
3, rue de Varembé
P.O. Box 131
CH – 1211 GENEVA 20
Switzerland
Phone: +41 22 919 02 11
Fax: +41 22 919 03 00
E-Mail: info@iec.ch
www.iec.ch

**Underwriters Laboratories (UL):**
www.ul.com

**Lloyd's Register:**
www.lr.org

**American Society for Testing and Materials (ASTM):**
www.astm.org
ASTM International,
100 Barr Harbor Drive,
PO Box C700,
West Conshohocken,
PA, 19428-2959 / USA
Phone:+1 (610) 832-9585
Fax: +1 (610) 832-9555

**International Maritime Organisation (IMO)**
www.imo.org
Visitors' Guide to the IMO Library

**American Standards Institute (ANSI):**
www.ansi.org
Customer Service Department
25 W 43rd Street, 4th Floor
New York, NY, 10036 / USA
E-Mail: ansionline@ansi.org
Phone: +1 212-642-4980
Fax: 212-302-1286

**DKE**
Deutsche Kommission
Elektrotechnik Elektronik
Informationstechnik im DIN und VDE
Stresemannallee 15
60596 Frankfurt am Main
Germany
Tel.: +49 (0) 69 / 63 08-0
Fax: +49 (0) 69 / 6 31 29 25
E-Mail: dke@vde.com
www.dke.de

**CENELEC (European Committee for Electrotechnical Standardization):**
www.cenelec.org
For any information contact the CENELEC Online Info Service:
E-Mail: Info@cenelec.org
Phone: Int + 32 2 519 68 71

# Index

## A

accident prevention   63, 64, 67, 69
additional equipotential bonding   79
all exposed-conductive-parts   41
AMP   89, 91, 95, 145
ASTM   81, 84, 130, 133
automatic disconnection   20, 21, 23, 36, 37, 39, 42–44, 55, 69, 169, 170, 176
auxiliary circuit   65, 78

## B

body resistance   67

## C

conductor   163
coupling   60, 97, 109, 132, 135, 136, 138, 139, 141, 146, 158
current protective device   37
current-carrying capacity   57, 76

## D

direct contact   36, 67, 77, 103, 113
disconnecting time   169
double fault   55, 148
double-faults   166

## E

earth fault   43, 49, 60, 65, 70, 76–79, 89, 91, 106, 110, 127, 129, 148, 177
earth fault current   51, 165, 166
earth fault monitor   135, 142
earth fault relays   110, 127, 142
earth fault current $I_d$   161, 162
earth fault lock-out device   80
electrical accidents   85, 104, 116, 119, 120
electro accidents   113, 122–124
emergency power supply   90
equipotential bonding   21, 23, 48, 49, 50
equipotential bonding conductors   54
esidual current monitor (RCM)   96
esidual current protective device (RCD)   39
exposed-conductive-part   25, 26, 47, 68, 79, 169–171
extra-low voltage   17, 36, 46, 78
Extra-Low Voltage (FELV)   44
extraneous conductive part   38, 47, 53
extraneous d. c. voltage   126, 139, 179

## F

fatal accidents   104, 122, 123
fault indication   97, 132
fire safety   63–65, 69, 75
first fault   38, 42, 43, 55, 58, 66–69, 169, 170
first insulation fault   39, 49, 82, 108, 110, 161, 169, 170
first short-circuit
    to exposed-conductive parts   110
fuse protection   60, 61

## I

IMD   42, 59, 60, 77, 110, 125, 127–132, 155, 156, 174

195

indication 20, 60, 64, 69, 78, 79, 82, 86, 89, 91, 103, 141, 143, 153, 155, 159, 174
indirect contact 36, 55, 86, 108, 110, 113
influence quantities 107, 110
inspection 54, 104
insulation fault detection 153
insulation measurement 108, 141
insulation monitoring 49–51, 58, 60, 64, 77, 81, 84, 86–89, 91, 98, 104, 109, 131, 135, 145, 146, 163, 173
insulation monitoring system 49, 87, 125
insulation resistance 49, 50, 55, 60, 65, 69, 70, 71, 78, 79, 82, 87, 89, 91, 98, 103–110, 125, 127, 130–132, 135–137, 142, 146, 148, 156, 159, 161, 162, 165, 170, 174, 175, 179, 182
internal resistance 60, 67, 77, 100, 126, 136, 142, 161, 181
isolated power system 174
isolating transformer 85, 135

## L

leakage capacitance 66, 106, 149, 156–160, 162
leakage current 42, 49, 67, 68, 94, 144, 149, 150, 157, 158, 175
leakage current measurement 160
leakage impedance 138, 159, 161

## M

maximum disconnection times 38, 39
monitoring device 33

## N

nominal voltage 44, 46, 63, 65, 77, 82, 90, 110, 125, 126, 131, 156

## O

ohmic insulation resistance 145, 155
operating safety 63, 64, 69, 87, 148, 152
overcurrent protective 40
overcurrent protective device 41, 42, 53, 59, 148, 176

## P

pathophysiological effects 42, 121, 169
patient environment 174
PE conductor 40, 141, 150, 158
PELV 21, 22, 44, 46, 55
periodic maintenance 90
phase-to-earth capacitance 162, 164
phase-to-phase 162
phase-to-phase voltage 128, 160, 161, 166
physiological effects 115, 121, 176
predictive maintenance 72
preventive insulation measurements 104
preventive maintenance 64, 69, 71, 94, 96, 141, 171
protective conductor 23, 27–30, 38–44, 47, 49, 50, 53, 54, 60, 68, 78, 79, 87, 108, 110, 135, 152, 162, 176, 180
protective conductor system 77
protective device 26, 31, 32, 38, 40–43, 47, 49, 56–58, 61, 79, 96, 98, 108, 110, 144, 161

protective earthing   38, 67
protective equipotential bonding   20, 23, 37, 38
protective measures   20–23, 35, 36, 44, 47, 48, 58, 68, 69, 76, 77, 79, 87, 94, 97, 98, 103, 106, 108, 110, 122, 125
pulse superimposition   141, 143

# R

remote indication   129
remote maintenance   72
required insulation values   90, 156
residual current monitor (RCM)   37, 109, 131
residual current monitoring   103, 108, 131
residual current monitoring device (RCM)   42
residual current protective device (RCD)   40, 42, 47, 58, 71, 96, 103, 108
resistance to earth   110, 135
response time   110, 111, 126, 141, 156, 180
response value   78, 87, 90, 110, 126, 136, 143, 146, 148, 150, 155, 156, 179, 180

# S

second fault   39, 43, 49, 55, 58, 169, 176
SELV   21, 22, 44, 46, 55
short-circuit   56–61, 66, 78, 79, 107, 161, 168, 175
short-circuit current   51
short-circuit protection   60, 86

short circuit to earth   107, 162
short-circuit to exposed conductive part   49, 51, 53, 65, 107, 170, 171
superimposed voltage   138
supplementary equipotential   38
supplementary protective equipotential bonding   50, 51, 53, 54, 110, 111
system leakage capacitance   67, 91, 110, 136, 138, 141, 146, 157, 180

# T

testing   54, 71, 78, 93, 100, 103, 105, 106, 108, 124, 125
TN systems   27, 31, 38–40, 43, 55, 56, 68, 82, 103, 104, 176
total earthing impedance   42, 68
touch voltage   50, 67, 77, 94, 107, 114–117, 119, 121, 122, 176
TT systems   32, 38, 41, 44, 55, 82, 107–109, 144, 148, 175
types of system earthing   25, 26, 129

# U

underground mining   63, 75–77, 79
unearthed IT systems   35, 49, 65, 66, 82, 87, 110, 153, 168

# V

voltage asymmetry   127, 142
voltage ratio   161

# Z

zero-crossing   165–168

# Technical Dictionaries

Gundlach, H.

**Elektrotechnik in drei Weltsprachen Deutsch – Englisch – Spanisch**

Ein Wörterbuch mit Erklärungstexten zur Elektrotechnik sowie auf CD-ROM das Programm GLOSARIO zur automatischen Rohübersetzung, zusätzlich mit vielen Abbildungen, Tabellen, Grafiken und Thai als vierter Sprache

2003, 340 pages, DIN A5, hardcover
ISBN 3-8007-2639-4, with CD-ROM
**43.00 €** / 71.50 CHF*

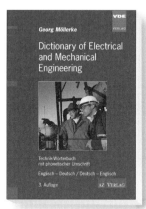

Möllerke, G.

**Dictionary of Electrical and Mechanical Engineering**

Technik-Wörterbuch
mit phonetischer Umschrift
Englisch-Deutsch
Deutsch-Englisch

3rd ed. 2004, 231 pages
12 cm x 17,5 cm, paperback
ISBN 3-8007-2782-X
**18.00 €** / 31.90 CHF

**VDE VERLAG GMBH** · Berlin · Offenbach
Bismarckstraße 33 · 10625 Berlin · Germany
Phone: +49/30/34 80 01-0 · Fax: +49/30/341 70 93
E-Mail: vertrieb@vde-verlag.de · **www.vde-verlag.de**